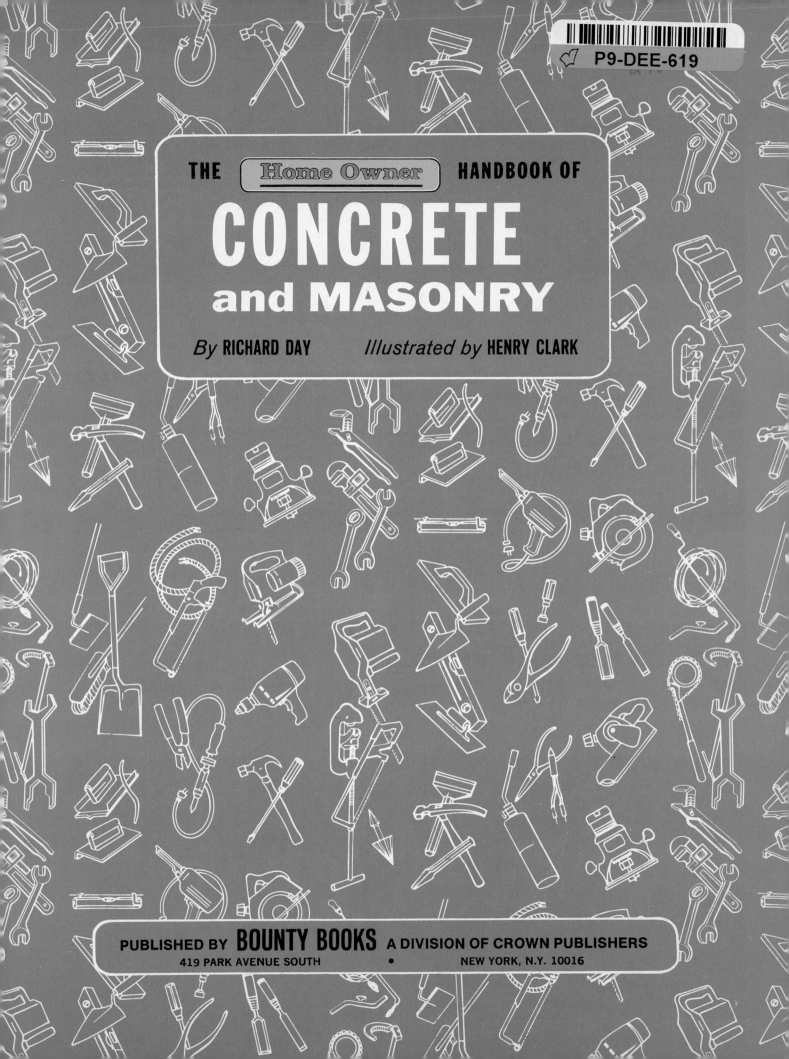

THE **Home Owner** HANDBOOK OF

CONCRETE
and MASONRY

By RICHARD DAY   *Illustrated by* HENRY CLARK

PUBLISHED BY **BOUNTY BOOKS** A DIVISION OF CROWN PUBLISHERS
419 PARK AVENUE SOUTH   •   NEW YORK, N.Y. 10016

LARRY EISINGER: *President/Editor-in-Chief*

ROBERT BRIGHTMAN: *Editor* ● JACQUELINE BARNES: *Associate Editor*

L. E. MARSH: *Art Director* ● JOHN CERVASIO: *Art Editor*

HOWARD KATZ: *Production*

Special Photography by FRED REGAN

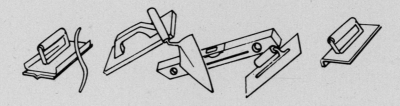

ACKNOWLEDGMENTS

We gratefully acknowledge the help of the following firms:

Baxter Concrete Products
Brick Institute of America
Cement and Concrete Association, England
Escondido Cement Products Co.
Hazard Products, Inc.
National Concrete Masonry Association

Portland Cement Association
Sakrete, Inc.
Standard Dry Wall Products
Superior Fireplace Company
Vulcan Waterproofing Company

Created by EISINGER PUBLICATIONS, Inc.
233 Spring Street, New York, N.Y. 10013

CONTENTS

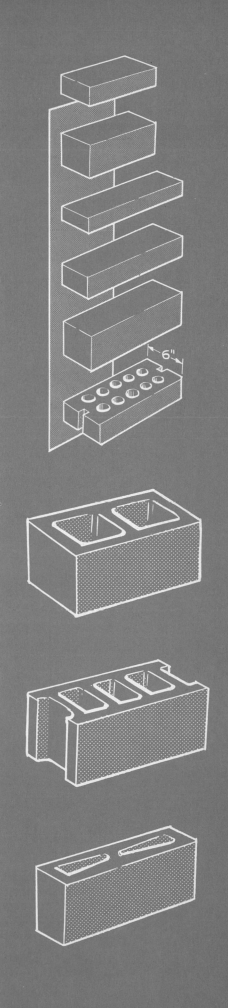

All it takes to create beauty in masonry is some wise planning. Stone, brick and concrete grille-block walls, as well as patio and unusual firepit, were all built with sacked concrete products in easy stages.

The Basics of Concrete

What you should know about concrete, the most universally used building material in America.

CONCRETE IS THE WORLD'S most useful low-cost building material. It builds roads and streets, airport runways, office structures, dams. Around your house, concrete will make a sidewalk, driveway, patio, pool deck, floors, walls, footings and a host of other projects. You can either hire a concrete contractor to make these things or do it yourself. Do-it-yourself saves money, and working with concrete is not hard, if you tackle each project in easy-do stages. In fact, working with concrete is fun!

Even if you have no experience with concrete, you can make durable, good-looking slabs and walls the first time. What you need to know is contained in the following chapters.

CONCRETE vs. CEMENT

First understand the difference between concrete and cement. Many people use the two words interchangeably, but they're not. Concrete is to cement as a cake is to flour. Concrete is a mixture of ingredients that includes what is called portland cement but contains other ingredients too. Concrete contains portland cement, sand, stones (the term we'll use to mean gravel or crushed stones) and water.

The most important ingredient in the mix is the portland cement, which is manufactured cement made by firing finely crushed rock of the correct composition in huge kilns. On completion it is packaged 94 pounds to the bag and sold by building materials dealers. One bag holds one cubic foot of cement.

Cement, with water added, is used as a "glue" to hold the sand and stones in a concrete mix tightly together. The whole mass hardens by what is called *hydration* into something like rock. Unlike most glues though, the process needs no air. Concrete can harden even underwater.

A good concrete mix must contain enough portland cement so that each particle of sand and stone is coated with the cement-water paste. If too much sand is present, the mix is weakened. A good concrete sand is coarse, containing particles from about 1/16 inch on down to dust-sized ones. The stones used in home projects are usually ¾ or one inch maximum size. They contain stones from that size down to stones as small as the largest sand particles. Thus, with the sand and stones mixed together the concrete contains a full range of sizes from the largest to the smallest.

This makes for an efficient mix. If any sizes are either missing or in short supply, the mix requires more cement paste and is therefore weaker.

Usually about two cubic feet of sand and 2½ cubic feet of gravel combine well with a one cubic foot (a sack) of cement to make about 4½ cubic feet of concrete. These figures don't compute; that is, 1 plus 2 plus 2½ don't equal 4½. The reason is that the smaller particles fit in among some of the larger ones taking up space that was empty.

LIQUID AND SOLID

Concrete comes in two states or forms: the

Precast solid concrete slabs 1½" thick set in sand make a stylish patio. Stones fill spaces between tiles. These Spanish-motif slabs were poured in a rented form, took just 20 minutes each to make.

Right. Once away from thinking of concrete as plain, smooth-troweled and gray, you can do great things with it. Combined with clay tiles, it makes a striking entry walk. Joints are tooled.

plastic, or wet state, just after its ingredients are combined, and the hardened state after the hydration process has taken place. In its plastic stage concrete can be shaped by pouring it into forms or finished by tooling the exposed surface, or both. While concrete is plastic you can do pretty much anything you want with it, limited only by the time you have before it sets up too hard to work. Some of the most pleasing and useful shapes and finishes for concrete take the least skill to produce.

TO MIX OR NOT

You can proportion the various concrete ingredients yourself and mix them or you can get them already mixed in several forms. Which way you do it depends on how much money you can spend vs. how much work you want to do yourself. Mix-it-yourself has the advantage of lower cost. Besides that, you make only as much mix as you can effectively handle. It does, however, involve you in buying and transporting the ingredients, measuring them out and shoveling them into a mixer, or else mixing them in a pile on the ground. Since this is real hard work, hand mix-it-yourself should be limited to smaller projects. The larger ones can be handled by renting or buying a concrete mixer.

All of the tool rental shops carry concrete mixers, the per day rental cost of which varies with the capacity and if the unit is gas or electric powered. Three or four cubic foot mixers are very popular and cost about $100-$120. When kept clean they last a lifetime and, surprisingly, have a good resale value. Make it a point to check the local classified ad section of your newspaper; mixers always seem to be found in this kind of advertising space.

Mix-your-own from *ready packaged* gravel or sand mix requires only the prescribed amount of water shown on the bag. This is a handy way to get concrete but it is more expensive than if you purchased the cement, sand and stone and mixed all the ingredients yourself. Ready packaged mixes are blended mechanically at the factory and generally assure you of getting a stronger mix because the ingredients are thoroughly mixed. Since we are always trying to save dollars, apply your basic math to figure out the cost differential between the quantity you require if supplied with ready packaged mix as against the cost of separate cement, sand and gravel.

Big projects—a driveway, walks, footings or a full-sized patio—call for ready-mixed concrete which comes delivered in a big truck-mixer.

Is your sand clean enough to make good concrete? To check put some sand in a glass, add water and shake. Next day, the silt layer should measure less than 1/8 inch for a 2-inch sample.

The cost is about 50 per cent more than buying separate ingredients. The drawback of ready-mixed concrete is that you usually must order at least three cubic yards of mix.

Ready-mix trucks have a capacity of anywhere from six to nine yards although in some areas of the country small two- or three-yard trucks are serving do-it-yourselfers. All of you have seen ready-mix concrete trucks—they are so large who could possibly miss them? But one new type of ready-mix truck that produces *metered concrete* is gaining popularity, especially in areas where home owners and small contractors require anywhere from one to three yards of concrete. This type truck, with a total capacity of nine yards, works on a different principle than the usual drum mixer where all the contents are mixed at one time. Rather, the amount of concrete required is dialed, and the prescribed amount of cement, sand, gravel and water is released and *mixed in the pouring trough* via screw action! The major advantage is the owner gets the exact quantity of concrete, but another equally important advantage is that the concrete is absolutely fresh! Thus, even with a five-yard pour the concrete can be mixed in smaller batches if you cannot handle a full five-yard or larger quantity efficiently. Check your local ready-mix firms for this type of "metered" equipment or contact the manufacturer, Concrete Mobile in Lancaster, Pennsylvania for a nearby firm using their equipment.

If you plan to do much concrete-masonry work around the house, buy your own electric or gas mixer. It should last a lifetime and lets you easily handle projects of less than 3 cu.yd.

When you mix your own concrete, you're responsible for selecting, proportioning, mixing and transporting the ingredients to the casting site. Having a wheelbarrow is absolutely necessary.

When you use ready-mix, the producer takes care of preparing the mix. All you must do is place and finish it. It helps to bring the truck-mixer in close enough to dump right on the site. Here part of a fence was removed to let the mixer back into a patio jobsite and dump directly on the sub-base. Ordinarily, the ready-mix driver is not allowed to help you with the pour.

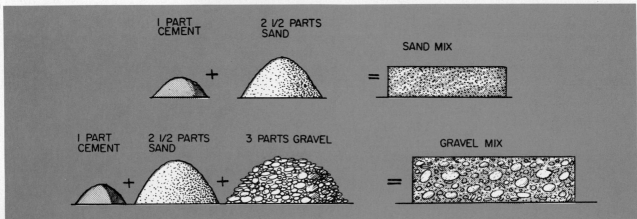

Tools and Techniques

The tools for working with concrete are basic hand tools; most likely you have them in your basement or garage.

TOOLS FOR CONCRETE and masonry work are not complicated nor are they expensive. You probably have most of the basic tools because almost anything you do in building—including concrete and masonry work—calls for hammer, saw, square, chalkline, tape measure or rule and a level-plumb combination. You'll need them to build forms, lay out footings, measure for quantities, etc.

For finishing concrete you start with what is called a *strikeoff*. It's simply a straightedge. Most often the strikeoff is nothing more than a straight 2x4 long enough to reach across the tops of the forms. Its purpose is to cut the concrete off level with the forms leaving no high spots and no low spots.

On projects up to about six feet wide, one man can work the strikeoff. Wider ones call for two men, one on each end. The strikeoff is worked by see-sawing it back and forth and with each stroke pulling it farther along. It's tilted toward the direction of travel to give it a cutting edge. Low spots should be filled in ahead of the strikeoff so that a little concrete wells up ahead of the edge. After covering about 30 inches in the first pass, a second pass is made to knock off any high spots that are left and to fill-in any low spots. On the second pass the straightedge is tilted backward for a filling rather than a cutting action.

INITIAL FLOATING

Right after strikeoff comes the use of either a *bull float* or a *darby*. You can make these tools, too, as shown in the drawing. The bull float, with its long handle, is best for reaching in where the darby cannot. Otherwise the darby is better. Either tool is designed to fill any voids left by the strikeoff and to push stones slightly below the surface where they're out of the way of further finishing operations.

A bull float is a long-handled, large-bladed float made of wood or metal. Wood is recom-

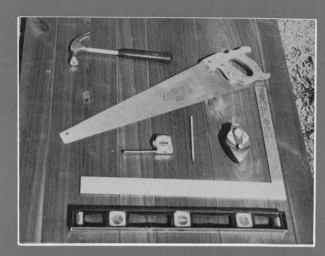

Hammer, saw, tape, pencil, chalkline, square, level are basic tools (above). Finishing tools include sponge float, edger, steel and magnesium trowels and wood float (below).

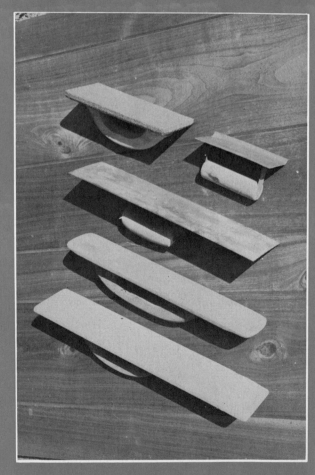

10

mended. As the bull float is pushed out over the surface, its handle is lowered to raise the top and prevent digging in. On the return stroke the handle is positioned to make the blade flat on the surface so it can cut off bumps and fill in holes. If there isn't enough concrete on the surface to fill in all the holes, some should be shoveled into the low spots and bull floating repeated.

A darby is three to six feet long made of wood or metal. Wood is most popular. The handle is best placed low on the tool, as shown in the drawing, to give the most precise control. On its first pass the darby is held with the blade flat to the surface and see-sawed from right to left and back again to slice off lumps and fill in low spots. Then when the surface is level, the darby is swung in arcs one way, then the other, always with the leading edge raised. This fills in small holes left by the sawing operation.

When bull floating, darbying or leveling,

shape and smooth the surface and also work a little cement paste into the surface. Just-placed concrete can easily be overworked, so finishing should be stopped after the floating passes just described. A popular tamping tool called a *jitterbug* should not be used except on very stiff concrete! You don't need one.

EDGING AND JOINTING

Wait until all surface bleed water has disappeared before starting the rest of the finishing. How long this takes depends on wind, humidity, sun, and temperature. On a hot, dry, windy day with no shade, this initial set can come within minutes and on a cool, damp or cloudy day, it can take several hours. With air-entrained concrete, which doesn't bleed, you can begin finishing almost right away, as soon as the water sheen (if there is one) has disap-

One tool you can make yourself is the straightedge. Here a straight 2 x 4 is see-sawed along "wet screeds" created by finisher Jerry Woods on a sloped concrete garage floor. Screeds are later removed.

peared and the concrete can hold your weight on one foot with no more than a slight depression.

The first finishing step is to edge the slab all around. Done with an edging tool, this breaks the sharp corner to prevent later edge chipping. You can get edging tools in various curves, but most acceptable is the ¼-inch radius. They come in large-blade (about 10x6-inch) and the more popular small-blade, curved-end edgers (about 6x3-inch). Material is steel, stainless steel or bronze. Insert the bit of the edger downward into the joint between the concrete and its form and stroke it back and forth until the edge has been shaped, then move forward to the next portion. Flat edgers should be lifted on the leading edge to keep them from digging in. Try not to let the edger leave too much of an impression in the surface that may be tough to remove later. Usually edging is done just before each finishing operation, even the final one.

A jointing tool is like an edger except that the bit is in the middle and it leaves a groove with curves on both sides. The idea of jointing is to divide a slab into panels small enough so that when the concrete shrinks on drying—shrinkage cannot be avoided—the cracks that form will fall conveniently in the joints. These joints are called *control joints*. To make them work, the bit of the jointing tool must be at least 1/5th the depth of the slab Thus, for a four-inch slab, a ¾-inch bit is minimum. Any less and the tool is called a *groover*. Its grooves would be for appearance only, not for control of cracking. Cracks likely will then occur at random in and out of the grooves.

Most dealers sell groovers as jointing tools and you'll need to insist on a proper bit depth.

Jointing is done like edging, but to get a straight joint, a board is laid down on the surface and the tool guided against one edge of it. Subsequent jointing will follow the groove without a board.

Float and trowel is the final finishing process. For the first troweling, the surface should be floated (left). Follow immediately with steel-troweling (right). Slapping will draw up water.

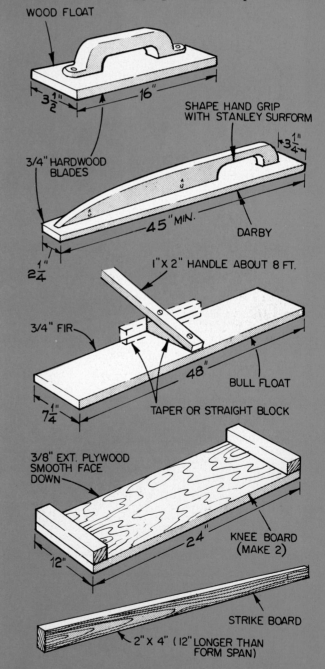

WOOD FLOAT

3½"　16"

¾" HARDWOOD BLADES

2¼"

SHAPE HAND GRIP WITH STANLEY SURFORM

3¼"

45" MIN.

DARBY

1" X 2" HANDLE ABOUT 8 FT.

¾" FIR

48"

BULL FLOAT

TAPER OR STRAIGHT BLOCK

7¼"

3/8" EXT. PLYWOOD SMOOTH FACE DOWN

24"

12"

KNEE BOARD (MAKE 2)

STRIKE BOARD

2" X 4" (12" LONGER THAN FORM SPAN)

Darbying, as shown on this page, should follow right after strikeoff. It is done in two steps to smooth the surface and prepare it for floating. The home-made darby above is see-sawed over the fresh concrete surface to slice off high spots and fill in depressions left by the straightedge. In this manner, it is used much like a straightedge. Fashion your darby from 1 x 2s and 2 x 2s so it is both light and easily used.

Next, the darby is swung in large arcs over the previously darbied surface to further smooth it prior to floating. When the slab is large, generally the unreachable center, portions are bull-floated rather than darbied. Darbying and bull-floating with long-bladed tools help to eliminate "birdbaths" from being formed in the soft surface. Once you can use a darby and bull float all your future jobs will be easier.

FLOATING

Right after edging and jointing, floating further embeds large stones beneath the surface and removes any roughness left by previous finishing. It also compacts surface mortar for a tough, hard-wearing layer readying it for the steps to follow. A float may be wood or metal; you can make the wood one yourself. When finishing air-entrained concrete, a magnesium float is best. A wood one tends to tear the surface. Another advantage of a magnesium float is that it slides easily over the concrete.

To use a float, hold it flat on the surface and swing it in sweeping arcs one way and the other. Floating puts an even but not smooth texture on concrete. If you want a surface with good slip-resistance, floating may be the final finishing operation. The finish left by a wood float is rather coarse. That left by a magnesium float would be fine for a sidewalk, driveway, or pool deck.

TROWELING

Hand-troweling immediately follows floating. In fact, both tools are carried and one is used for support while reaching out with the other. The indentation left is smoothed over before leaving that area.

A hand trowel is a fine piece of steel, even stainless steel, that puts a smooth, dense surface on the concrete. Pros make use of two trowels, a larger 18x5-inch one for first troweling and a smaller 12x3-inch one for final troweling. To save money, you can get along with just one, an in-between 14x4-inch trowel.

For initial troweling hold the trowel flat on the surface. Tilting it makes ripples that tend to show up in the finished slab. Move the trowel in sweeping arcs letting one arc smooth out the edge marks of the one before it by lapping halfway. Pick out any stones that intrude into the surface and toss them away.

First troweling produces a good surface for most uses, but for a really easy-to-clean surface, as in a basement, where a nonslip footing isn't as important as it is outdoors, you can trowel several times. The time for the next troweling is when pressing your hand onto the surface makes only a slight impression. On second troweling hold the trowel blade with its leading edge up slightly to further compact the concrete. If you give a third troweling for an almost glossy surface, wait until the concrete has gained additional strength. Hold an even greater tilt on the trowel during this pass. The

14

If you want your control joints to be square with the sidewalk, driveway or whatever you're building, form them along a line snapped between form marks. Control joints are at 8'-20' intervals.

trowel should then make a ringing sound as you draw it over the surface.

KNEE-BOARDS

To get out onto a large concrete slab and finish it, you'll need a set of knee-boards. Make them yourself, as shown in the drawing, of Masonite or plywood. To use them, place one in front of the other and step out into the center of the slab. Lay the knee-boards down about two feet apart so that your knees can rest on one board and your toes on the other. Finish the area in front of you, then move sidewise to the next area. Skip any spots that can be reached from outside the slab. Always work backwards after a sidewise pass so that impressions of the knee-boards are finished over. When you have completed everything unreachable from off the slab, put the knee-boards aside and finish around the outside of the slab. Keep your knee-boards. They're a permanent part of your concrete tool kit.

OTHER FINISHING METHODS

If you look over the displays at your building materials dealer's, you'll spot a number of different concrete-working tools that look as though they have the same purpose. One is a

Since the first floating and troweling is a combined operation, take both tools with you. If they are strong, you can rest your hand on one while using the other. Redwood floats are great.

sponge rubber float. It looks like a dense sponge glued to a wood or metal float. It is used to create a textured, slip-free surface on slabs after magnesium-floating.

Other interesting-looking trowels come two to a set. They're for finishing steps. One finishes the inside corner of the steps, the other finishes the outside corner. It's about the only way to get a complete finish on concrete steps.

Broom-finishes that are put on after final troweling can be fairly fine or coarse. To broom a finish, first dampen the broom, then pull it lightly over the surface. Test in a spot to see if you're getting the effect you want. Early brooming and stiff bristles make for coarser finishes; late brooming and finer bristles make for smoother finishes. The broom may be swirled or zig-zagged during its passes for a wavy finish rather than straight lines.

The mason's trowel is shaped somewhat like the palm of your hand but larger. The handle is wood and the blade is steel. Mason's trowels are made in various patterns (shapes) with narrow heel (4-7/8 inch) or wide heel (5-5/8 inch) and in various balances. Choose one that feels comfortable when you hold it. Weekend masons shouldn't get too large a trowel because they tend to tire the wrist and arm muscles.

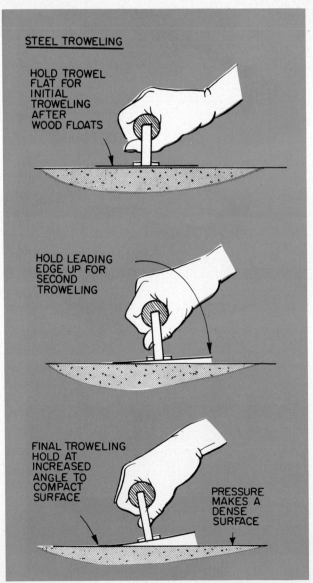

STEEL TROWELING

HOLD TROWEL FLAT FOR INITIAL TROWELING AFTER WOOD FLOATS

HOLD LEADING EDGE UP FOR SECOND TROWELING

FINAL TROWELING HOLD AT INCREASED ANGLE TO COMPACT SURFACE

PRESSURE MAKES A DENSE SURFACE

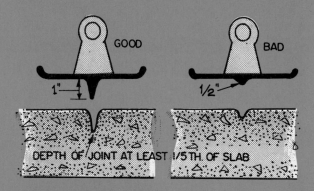

GOOD   BAD

1"   1/2"

DEPTH OF JOINT AT LEAST 1/5TH. OF SLAB

Bull-floating a slab lets you remove the high and low spots left by strikeoff without getting onto the fresh concrete. The long handle reaches to the center of just about anything you'd build.

Whenever you get a spare minute during a concrete-finishing project, use the time to clean tools that you're through with. A garden hose fitted with trigger nozzle will do the cleaning job.

Control joints should be formed with a jointer having a bit at least 1/5th of the slab's depth. Use a board as a guide the first time, then tool will follow the joint during steel troweling step.

LEVEL

A mason's level is longer than a carpenter's level, usually four feet long. You needn't buy a long level unless you plan on doing a great deal of masonry work. The greater length lets the level be used as a long straightedge for aligning courses of brick and block.

STRINGLINE

Masonry units are laid straight to a taut stringline. You can use nylon fishing line, if you wish, or buy nylon mason's line.

Handy to have with your line is a pair of line blocks. These attach to the masonry corners and let you stretch the line without inserting nails in the joints. You can get line blocks in wood, metal, or plastic.

MASON'S HAMMER

A carpenter's hammer is not much good for cutting bricks and concrete blocks. For that you'll need a mason's hammer. It has a square head at one end for chipping off pieces and a

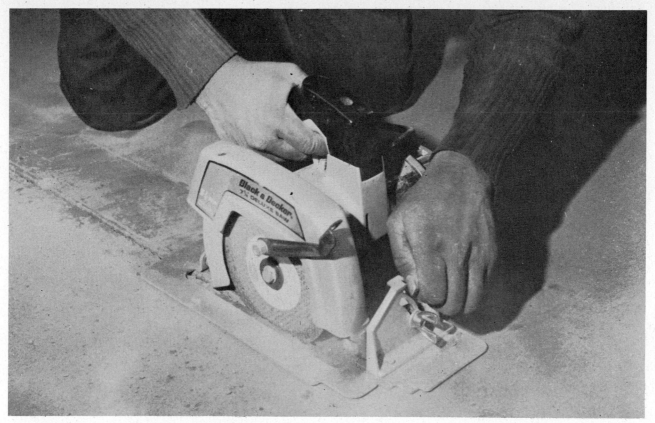

Another way of cutting the all important deep control joints is with an abrasive wheel mounted in a power saw. The advantage of this is that jointing can be left until after the slab has set hard enough to walk on. Don't wait too long, however, or the contraction will cause random cracking before you cut the joints to control it. Sawing too soon tears out stones. Most of our concrete highways use sawed joints.

chisel point at the other end for chipping lines across masonry units in cutting them. A 20-ounce head is a good size.

RULE

Some means of measuring the height of each course of brick or block is needed. You can either make a wooden pole—called a story pole —with the course levels marked on it, or buy a special mason's rule. These are available in folding wood or steel tape with the various-dimensioned courses clearly marked.

JOINTING TOOLS

The mortar joints in masonry work should be tooled to compact and smooth them. What you use for this will determine how the joints look. V-shaped tools make V-shaped joints. Rounded tools make concave joints. Special raker tools are made for raked joints. For concrete blockwork, the jointing tool should be about 20 inches long. For brickwork only, it may be much shorter.

MISCELLANEOUS TOOLS

In addition to the above masonry tools, you'll have use for a wire brush for cleaning off mortar. If you are to acid-clean your brick or block project to remove all traces of mortar stains, you'll need a pair of waterproof gloves. If you like working with gloves, a pair of vinyl- or rubber-dipped work gloves will save your hands while handling mortar and masonry units. If kept dry, leather gloves are more comfortable because they "breathe." They need not be full leather, just leather-faced for hard wear. Cloth work gloves containing small plastic gripper dots are tough-wearing, yet they are comfortable. The full-dipped gloves tend to make your hands sweat.

A brick chisel, a flat one with a wide face and perhaps 2½ inches across, may be useful for cutting bricks to exact lengths. It is laid across the brick and struck with a mason's hammer for a clean cut.

You may need a mortar box, hoe, shovel pail, garden hose, wheelbarrow, and some other tools. Often, if you don't have these, you can make-do without them. They help the job go easier, though.

17

Mixing Concrete

Water is the most critical ingredient in concrete. The less you add, the stronger the concrete will be. Left mix is too wet. Mix at center is just right. Right, too dry.

Concrete can be mixed by hand or by power; but large batches should always be power mixed.

ANY PROJECT REQUIRING up to about three cubic yards is an excellent prospect for mix-your-own concrete.

You can mix by hand or with a concrete mixer. Easiest, of course, to use is the mixer. Home-type concrete mixers come in all sizes from a five-gallon pail to a five-cubic-foot and even larger. The smaller mixers are good for making stepping stones, laying flagstones and other limited-pour projects. All but the smallest mixers are available either electric or gasoline powered. The gasoline-powered mixers cost more but are handiest because they're independent of an electrical outlet.

If you plan quite a few concrete projects over the next few years, it will pay to buy your own mixer. They last, practically without trouble, for a lifetime. The handiest ones are mounted on wheels for easy transporting. Some are designed to be moved about like a wheelbarrow.

MIXER CAPACITY

The actual capacity of any concrete mixer

is 60 per cent of its drum size. Thus, a five-cubic-foot mixer will actually handle batches of about three cubic feet. This size is known as a half-bag mixer because it will handle a half-sack of portland cement plus the sand, gravel and water. This is a very convenient size because it lets you proportion half-bags of cement, avoiding further measuring. The identification plate on the mixer gives its capacity in cubic feet.

Concrete that is improperly proportioned, mixed, placed and finished can suffer all kinds of troubles from cracking to scaling and dusting. So do it right and you'll bypass these problems!

Start with good materials. Portland cement, though not a brand, is good in all brands. Use either Type 1 or 2. They are suitable for everything you will ever want to make. Type 3 cement is for high early strength. It hardens quickly to permit early removal of forms. While you could use it, you don't need it. Other types of cement are for highly specialized purposes.

You can also get types 1A and 2A. The *A* stands for *air-entraining*, that is, this cement

The amount of water you add to a mix depends on how much free water is in the sand being used to make it. To find out, pick up a fistful of sand, squeeze and release. Dry sand at left contains no free water, falls apart. Average sand (center) stays in a tight lump but leaves no water on the hand. Wet sand, as at the right leaves your hand wet. Its a good idea to cover your sand with plastic every night.

To judge sand-stone proportions in a fresh mix, dump same in a pile and pass a trowel or shovel over the top. If the surface shows stones and holes as at left, it contains too much gravel. If it trowels beautifully with no signs of stones, it has too much sand (center). If it smooths without tearing, as at right, but shows some stones, it's right. To know how to recognize your proportions is a must for good concrete.

when mixed, creates air bubbles in the mix. When trapped within the concrete, these microscopic bubbles act as tiny "relief valves" for pressures caused by freezing water within the concrete. Instead of popping off a chunk of the surface, during sub-freezing weather, the freezing water expands harmlessly into the bubbles.

All concrete exposed to freezing should be air-entrained. Air-entrained concrete costs no more. If you don't make it with types 1A or 2A portland cement, you can get an air-entraining agent and add it separately to your mix. Air-entraining agents are not widely available to the home owner so the best place to get it is from a ready-mix concrete dealer who is accustomed to supplying air-entrained mixes to his contractor customers. Take a gallon can along with you and for a few dollars he'll sell you a lifetime supply; but ask the dealer for advice on how much agent to add to each batch since different manufacturers require different proportions. If it's the most popular brand—*Darex*—use it at the rate of two tablespoons per bag of cement. A half-bag mix, then, would get one tablespoon

of *Darex*. One important point; hand-mixing is not vigorous enough to entrain any air, so you needn't expect to make air-entrained concrete by hand.

QUALITY AGGREGATES

Quality in aggregates (sand and stone) is important, too. You should use only sand and stone that are known to make good concrete. Often raw materials dug out of the ground have to be washed, crushed, regraded, and reblended to make good concrete aggregates. For this reason, it's a good idea to buy the materials from a ready-mixed concrete producer or from a reputable concrete products dealer.

One way to judge the cleanliness of sand is to put two inches of it in a jar, add water almost to the top and shake thoroughly. Let it stand overnight and measure the thickness of the top layer just beneath the water. If the silt layer is more than 1/8-inch thick, the sand is too dirty to make good concrete.

Another way to judge aggregates is to look at the piles. Pick some up. Do they contain

With a good-sized half-bag mixer it's easy to batch cement. Just slice the bags in two with a shovel and add half a bag to each batch.

Clean-up of tools and mixer after a pouring project is important. The best way is with a garden hose trigger nozzle. Squirt right after use.

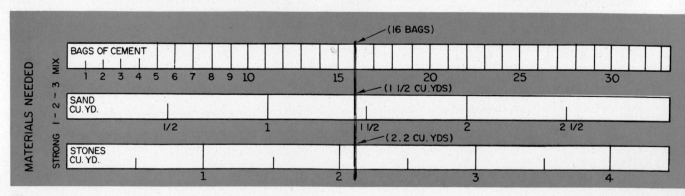

MATERIALS NEEDED — STRONG 1 - 2 - 3 MIX

BAGS OF CEMENT
(16 BAGS)
1 2 3 4 5 6 7 8 9 10   15   20   25   30

SAND CU. YD.
(1 1/2 CU.YDS)
1/2   1   1 1/2   2   2 1/2

STONES CU. YD.
(2.2 CU.YDS)
1   2   3   4

20

mostly one-size particles, or are all sizes present as there should be? Fine sand—the kind you'd want for a child's sandbox—is not suitable for making concrete. It contains few larger, coarse particles and tends to be all one-sized grains. Good concrete sand is coarse, sharp and grainy. In different locales it goes by different names, but if you ask for *concrete sand*, you should get the correct product. What you *don't* want for making concrete is masonry sand, sandbox sand or beach sand.

Water quality is important. The rule is, if you can drink the water, it's okay for concrete.

When you have all the materials on hand, you are set to mix and pour. In addition, forms, tools, helpers—everything else you'll need to complete the job—should be on hand before you start.

Arrange the piles of sand and gravel handy for shoveling into the mixer. You may find it handy to place all the materials on one side of the mixer and leave the other side free for dumping and hauling. Hook up a garden hose, ideally one with a trigger nozzle both for mix water and for clean-up. This saves running to turn off the hose. Another way is to let the hose run into a 55-gallon drum with the top removed. Then you can dip buckets of water from the drum as needed.

The more helpers you can muster, the better. However, by yourself you can tackle everything if you plan on doing a little at a time. As a general rule, you can probably handle the mixing, placing, and finishing of one cubic yard in one day. Two workers can do twice as much. If you have three, put one man on the mixer, one hauling and spreading, and the most careful one, finishing.

PROPORTIONING INGREDIENTS

The accuracy you use in measuring out ingredients for your concrete is up to you. The more accurately you work, the better concrete you'll produce. The simplest but least accurate

method is by counting the shovels of each ingredient. Most accurate is putting each material into measure-marked cans and emptying these into the mixer. For most purposes, the shovel-method works fine. See that shovels of cement, sand and gravel are all about the same size. This isn't as easy as it sounds because damp sand and the cement both tend to pile up on the shovel, whereas gravel tends to roll off. Try filling a wheelbarrow with sand and with gravel to see how shovel quantities compare.

Two kinds of concrete mixes are shown in the tables and on the back cover. Choose the one you want according to what project you're making. The Strong Mix is best for making small units with thin cross-sections, projects such as patio tiles less than three inches thick. The Economical Mix is best for most everything else including footings, foundations, and slabs thicker than three inches. It will not develop the high strength of the Strong Mix, but it goes farther for the money because it contains more sand and stones.

Proportions are shown in the first column. For example, to make an Economical Mix, in a small mixer, shovel in 5 shovels of cement, 12½ shovels of sand, 20 heaping shovels of stones, and add the equivalent of 2½ parts of water.

When using a half-bag mixer, the correct amounts of materials are shown in the next column.

WATER WEAKENS

No matter what size mix you make, the proportion of water should be no more than six gallons of water per sack of cement. Too much water weakens the mix.

Since sand contains some free water, or is so dry that it absorbs some of the mix water, the amount of water you put into a mix depends on the condition of the sand, on the day you use it. Average sand is wet, that is, it contains enough water to wet each particle. If you squeeze a handful and release it, average sand is wet

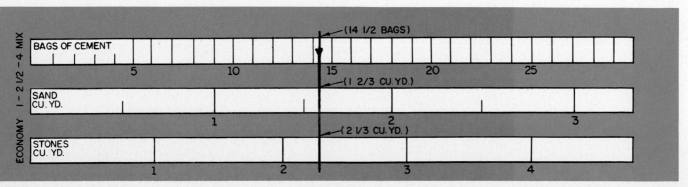

enough to stay in a tight lump. Test yours. If the squeeze-test leaves water on your hands, the sand is very wet and the mix water should be reduced, but only by about half a pint in a half-bag mix.

If, on the other hand, the sand is so dry that it falls apart when squeezed and released, mix water should be increased about half a pint over that shown for a half-bag mix.

The last column in each table shows how much cement, sand, and stones will be needed to make each cubic yard of concrete. When you order the materials, though, get about ten per cent extra to allow for waste. Estimating volume is described in a later chapter.

LOADING THE MIXER

Arrange to have the concrete mixer high enough to dump right into a wheelbarrow. A big rubber-tired contractor-type wheelbarrow is handiest. It holds about five cubic feet and gives a smooth ride without too much settling of the stones.

Load the mixer in this way: first put in about half the water, pouring down the sides of the rotating drum to clean it. Then shovel in half the stones and all of the sand. Let them

mix for half a minute, then add all of the cement and the rest of the stones. Also put in the air-entraining agent, if one is being used. Finally, pour in the rest of the water. A good way to measure the water is in a pail, filling to a depth mark taped or drawn on the side or inserting a stick with a depth mark on it. Just remember that water quantity should be consistent batch to batch or you'll find problems later on in placing and finishing of the alternate stiff and sloppy mixes.

Consider the first batch as a trial. If it's too dry, add more water and adjust the water mark accordingly. Too wet, lower the height of the water mark and save the trial batch by adding a little more cement, sand and stones to dry it up.

Quantities of sand and stones can be varied, too, to make an easy-working mix. The best way to judge is to let the mix churn for two minutes, then dump it into a wheelbarrow and examine it. Run the back of a shovel over the surface to smooth it down. Then compare with the photographs of good mix, stony mix and sandy mix. A good mix contains just enough sand and cement to make a closed surface without any runny cement-sand-water paste. A mix that is too stony is hard to work during finish-

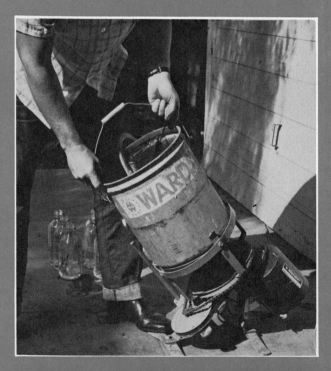

One of the handiest home-type concrete electric mixer made uses a 5-gallon pail for a drum. Ingredients for concrete or mortar are batched and mixed, then toted to site in same can.

Take care in measuring water. Good way is to find the correct amount for a batch, then mark a stick at the height it comes to in the batch can. Fill to mark each time, keep it just for water.

22

ing. One that's too sandy works well, but is not as durable as it could be.

One concrete mix, called sand-mix, is made without any stones. Use sand-mix concrete for thin sections of 1½ inches or less. It's made using the Strong Mix proportions, but without the stones. Yield, of course, is much less.

Mix concrete at least two minutes after adding the last ingredient but don't leave it in the mixer more than half an hour or it may begin to harden. When thoroughly mixed, the ingredients should fall off the mixer blades cleanly without the mix being sloppy.

Take these precautions when mixing. (1) Don't peer too closely into a rotating mixer drum. The slopping concrete can get you in the eye, a painful experience that will send you to the garden hose to flush out the debris. (2) Be careful when reaching into the turning drum with your hand or a shovel as when trying to retrieve a piece of cement sack or a twig. Shovels and arms can get broken that way. It's best to stop the mixer when removing things from it!

As soon as you dump the last batch of concrete for a project, put some water and several shovels of stones into the drum and let it revolve for 10 or 15 minutes while you place the last batch. This will clean the drum's insides. Likewise, if you stop for lunch, do the same thing. Finish the cleaning job with a hose while it turns. Also, remove cement from the outside of the mixer and from the wheelbarrow. Use a wire brush on stubborn deposits. Stop the drum and make sure no deposits are left in it. Also, when washing down equipment, don't let the cement-laden water run down a drain. It can harden and plug the drain.

HAND-MIXING

Mixing concrete by hand is a little different from machine-mixing. Dump the materials onto a smooth surface. Use a concrete slab or set down a sheet of plywood. First measure out the sand and stones into a heap. Add the cement to a crater formed in the center of the pile. Turn the dry mix until the whole pile is the same even color without streaks and lumps. Usually, turning several times from one pile to another does the job. Crater the heap again and start adding water a little at a time. Keep mixing until the whole pile is the same color and consistency. Adjust the ingredients, if need be, as described for machine-mixing.

Carry concrete from the mixer to the pour site in a rubber-tired wheelbarrow. The softer ride keeps the stones from settling to the bottom. Use wood planks as ramps. Its best to wear boots.

To make air-entrained concrete, add a teaspoon of Darex air-entraining agent to each half-bag batch as it mixes. Microscopic air bubbles freeze-protect the concrete. See your materials dealer.

Buying

There's more to picking up the phone and ordering mix. Here are some helpful tips we learned.

Getting a big truck-mixer in and out of your job is your responsibility. Here Carl Roth protected his driveway with sheets of old plywood before guiding driver in.

A PROJECT THAT REQUIRES three cubic yards of concrete or more is a good candidate for ready-mix. Ready-mix can be used on smaller projects, too. However, unless you can buy "metered" ready-mix in your area you will find most of the drum ready-mix producers make a minimum charge somewhere around three cubic yards, so you'll probably pay for that much whether you need it or not.

Ready-mix is purchased and delivered to your house just like a piece of furniture. All you do is call up and order it, and don't be surprised if you are scheduled for a specific time and date. The supplier takes care of selecting materials, proportioning them, mixing, delivery, and discharging the mix as nearly as possible to the place where you want it.

On the other hand, using ready-mix lets you in for plenty of hard labor. Having that much concrete to place and finish in a short time can be a big burden. Have plenty of help on hand, especially if the weather is hot and dry and the sun will be shining on your project. Concrete sets up quickly under such conditions. Two people can probably handle three to five cubic yards of ready-mix in a day. If all of you are new to concrete, it's better to have three. If

the mix has to be wheeled to the site, you'll need two more helpers.

You'll find ready-mix producers in the Yellow Pages of your telephone directory. Like other businesses, there are large dealers and small ones, good ones and not-so-good ones. It's hard for the occasional user of ready-mix to know the difference. Generally the big producers who serve many concrete contractors know the most about making good concrete. On the other hand, they are the least anxious to bother with a small load. It's practically impossible to get a Saturday delivery from one of them. If you do succeed, you'll probably find the price much higher on that overtime day. The small ready-mix producer may be a better bet for you. Choose your man on the basis of the service he offers. Will he bring the mix when you want it? Does he have a wheelbarrow to rent you? What is the farthest he can dump the mix from the rear end of the truck? If you need more distance, does he have an add-on chute?

ORDERING

Ready-mix properly ordered is half-way toward being good concrete. Here's how to do it:

Ready-Mix Concrete

Have enough strong help on hand before the ready-mix truck arrives. Once it begins dumping, things can get really hectic. Roth (right), Woods (center) and a neighbor help place the mix.

Ready-mix is indispensable for farm construction projects, which can involve quite a bit of concrete. Paving a barnyard or a patio is a good co-op project using many friendly neighbors.

Read your ready-mix producer this statement over the phone, filling in the blanks as you go:

"Can you bring me _____ cubic yards of ready-mix with one-inch maximum-size aggregate, a minimum cement content of six bags per cubic yard, a maximum slump of four inches and a 28-day compressive strength of at least 3500 pounds. It should also have (in freezing climates) an air entrained content of six per cent, plus or minus one per cent. And can you bring it at _____ o'clock on _____."

Your dealer should be impressed, because usually only engineers know enough to order ready-mix this way. It covers all the bases for quality.

Maximum aggregate size depends on the section thickness of your project. This should be as large as possible without being more than one quarter of the section dimension, (i.e., one inch aggregate for a four-inch slab). Slabs five and six inches thick can use 1½-inch aggregate. For building steps, a one-inch maximum size is preferred.

Slump is a measure of the workability (ease of placing and finishing) of the plastic concrete. Concrete with more than four inches slump begins to get sloppy and tends to be weak.

Much less slump and you'll have too tough a time placing and finishing.

Specifying compressive strength for ready-mix implies to the dealer that you are going to cast test cylinders on the job—very professional —to check up on the quality of his product. The minimum entrained air content ensures weather-resistant concrete while the air bubbles keep water from bleeding to the surface and getting in the way of your finishing operations.

PLANNING YOUR POUR

Be ready before the ready-mix truck arrives. Have your project formed and plan so the truck will be close to the forms so you can dump the mix directly into the forms. If you must wheel it in, have ramps ready. Put planks down where the truck will cross over a shallow buried sewer and lawn sprinkler pipes or soft ground. Protect thin-section sidewalks and driveways less than six inches thick from the weight of the truck-mixer by laying down planks or sheets of ¾-inch plywood. An overhead clearance of about 11 feet is needed for most trucks. Prop up wires, or plan how you will have the driver

There is no finer way to get ready-mix in place than with a concrete pump. The truck—or trailer-mounted contractor's pump can be positioned practically anywhere, just so the ready-mix truck can dump into its screened hopper. A 2" hose (foreground) leads to the pour-site. Pump takes the small-aggregate mix almost as fast as it can be dumped. You are lucky if you have a local concrete pumping contractor.

avoid them. Have your tools ready and your helpers on hand.

When the truck arrives, guide the driver into position for the first pour. See that he revs his engine up and mixes the batch for at least two minutes, then he can start dumping. The driver's job is to control the dumping, nothing else. Most drivers are not allowed to help, even if you are overworked and he hasn't much to do.

Should the mix prove too stiff for you to handle, you can have the driver add water at the rate of one gallon per cubic yard. He does it from a storage tank on the truck but remember that adding water will void the compressive-strength minimum that you have specified because *too much water weakens concrete.* Keep in mind that concrete should not be so wet that it flows into place. After adding water, the driver should run the drum at mixing speed for two full minutes before dumping any more concrete.

Most ready-mix producers allow you a specific amount of time per yard for unloading. If dumping the whole load takes longer than anticipated, you have to pay for the additional time at something like $12 an hour and up. Time starts when the driver arrives on the job and runs until he is ready to clean up and leave.

If your project requires more than one load of ready-mix, have the two trucks arrive about a half hour apart or have the same truck go back to the plant and get your second load. That gives you a breather between loads. Use it to spread and to begin finishing the concrete.

A good many communities have what is a real do-it-yourself innovation—trailered ready-mix. The best trailer-mixers contain a one-cubic-yard mixing drum. The trailer attaches to a hitch installation on your car. At the ready-mix plant, you pull the trailer under a chute or hopper and it is filled with mix. The self-contained mixer engine is started and it mixes while you drive home. Once there, you pull the trailer to the dumping site and empty the mixer. While you begin to place the mix, someone can drive back for the next load, if necessary, or clean and return the mixer to avoid a $10 extra charge.

Trailering is ideal for projects up to several cubic yards. The best place to rent a trailer-mixer is from a ready-mix producer, because he knows concrete. Cost is about the same as for ready-mix with 1½ hours of trailer time included. Overtime is figured at $6 to $10 an hour

PUMPED CONCRETE

If getting a ready-mix truck close to your job site is a problem, use a concrete pumping contractor's services, if available in your area. For $40 to $60 he will come out and place the mix for you using a two-inch hose connected to a truck- or trailer-mounted concrete pump. You simply position the ready-mix truck to dump into the hopper of his pump. He starts the pump and hands you the end of the hose. Concrete will flow steadily from it like toothpaste from a tube as long as you need it!

Using a concrete pumping contractor lets you place all the concrete for a large patio in about 15 minutes, by yourself. There is no additional truck-mixer time charge, and you needn't

pester your friends to help. The mix can even be pumped into a basement through an open window. Uphill or down, is no problem, within reasonable limits. When finished, the contractor cleans his own equipment, presents you the bill, and leaves when you pay. It's an ideal way to get ready-mix. Some ready-mix dealers have pumps. Others let you arrange for pumping separately, getting the ready-mix and pump on the job at the same time.

The only drawback to pumping is its cost. The minimum charge usually covers four to six cubic yards. More than that will bring an additional charge per yard of about $4 to $6.

Look for a pumping contractor under "Concrete Pumping Service" in the Yellow Pages. Ask the man you call whether he has a two-inch pump to handle small jobs. If he doesn't, the cost would be prohibitive. Try another contractor or ask your ready-mix producer to recommend one. Some areas have many, some have none.

If you use a pumping service, be sure to order your ready-mix with 3/8-inch maximum-size aggregate so it will go through the pump. Often, this is called a *grout and pump mix*. At any rate, your ready-mix producer should be told that you plan to pump-place the mix so he can specify it accordingly.

Pumped concrete comes from the hose end like toothpaste from a tube. Here concrete pumping contractor John Dolstra controls the hose, while Chuck and Randy Kearns help hold it in place.

It took just 20 minutes to pump the Kearns patio in place. As each square filled with mix, Dolstra moved to the next. All wheeling, shoveling, raking was avoided. Pumping costs, but is worth it.

How Much to Order

How to order or mix the concrete you need without running short or having a big left-over pile for which you have no use.

When that ready-mix truck backs up to your job, you cannot help but wonder whether you ordered enough to fill the forms. Remember, it's better to have a little too much than not enough to fill the forms.

EARLY IN THE BUILDING PROCESS you must figure out how much concrete you'll need. After you calculate the volume the job requires, round off the figures to the larger quarter-yard or half-yard, rather than down. If, you end up with a concrete requirement of 6.3 cubic yards, order 6½, not 6¼. A little extra concrete left over after the job may be used to cast step-stones (if you have a form handy) or it can be spread out in a predug hole as a free-form fish pond. One smart home owner uses left-over ready-mix to lengthen his backyard walk. He has the forms ready and dumps the excess between them each time he uses concrete. His walk grows a little every year.

For cast-in-place concrete slabs, figure the area covered by the slab. If it's square or rectangular, the job is easy. Simply multiply the length times the width, both in feet, and you have the area in square feet. If it's triangular and one of the corners is a right-angle (90° angle), find the area by multiplying the dimensions of the two sides that meet at the right angle, then take half that amount as the area (see drawing). If the area is triangular, but without a right angle, you can divide the triangle in two right-angle triangles by an imaginary line. Then figure the areas of these and add them together. Other shapes are estimated as shown in the drawing.

An irregular area, such as a kidney-shaped pool deck is figured differently. Sketch a miniature of it on squared paper, letting one-inch squares equal one-yard squares. First count up all the whole squares within the area and write that figure down. Then guess at what portion (in tenths) of each of the partial squares is part of the project (see illustration). Write them down. Add it all up and you have the approximate area in square yards, probably within tenths either way.

ESTIMATING VOLUME

Now that you know the area to be covered by a slab, you can figure its volume. We recommend starting in square feet. Then, since you want the volume in cubic yards, convert area to square yards. (If it's square feet, divide by nine to get square yards.) Then apply the thickness to find volume. Thickness is nearly always measured in inches, and an inch is 1/36th of a yard, so first multiply the area in square yards by the thickness in inches and then divide by 36 to end up with cubic yards.

If you order materials by weight rather than by volume, remember that sand weighs about 2,500 pounds per cubic yard and ¾-inch stones weigh about 2,800 pounds per cubic yard.

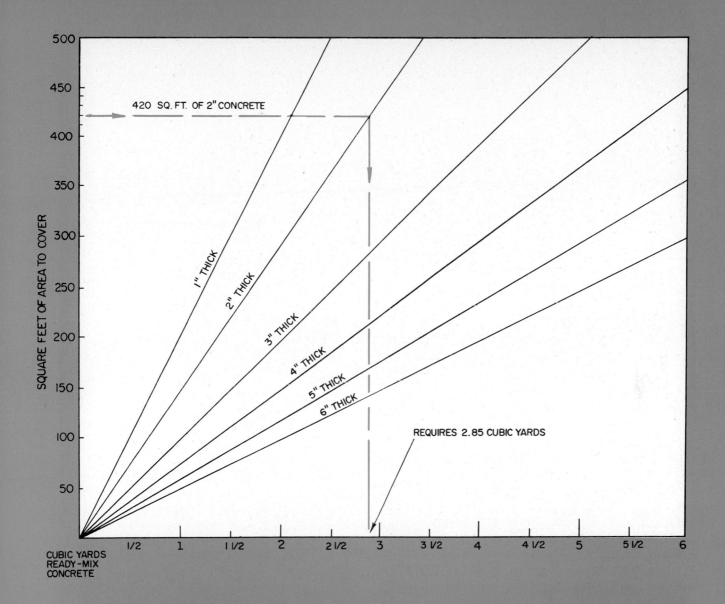

500

450

420 SQ. FT. OF 2" CONCRETE

400

350

SQUARE FEET OF AREA TO COVER

300

250

1" THICK

200

2" THICK

3" THICK

150

4" THICK

5" THICK

100

6" THICK

REQUIRES 2.85 CUBIC YARDS

50

CUBIC YARDS
READY-MIX
CONCRETE

1/2   1   1 1/2   2   2 1/2   3   3 1/2   4   4 1/2   5   5 1/2   6

HOW MUCH READY-MIX CONCRETE TO ORDER —   (INCLUDES 10 % FOR WASTE.)

The back-cover chart (duplicated above) simplifies figuring how much cement, sand, and stones you need to make the concrete you need for any slab. To use it, all you need know is the size of the slab in square feet and its thickness in inches. Quantities are shown for slabs up to 500 square feet and from one to six inches thick.

For example, suppose your slab measures 12 x 35 feet for a total of 420 square feet and that it's to be two inches thick. Start to the left opposite "Area to Cover" at the 420 square feet level and draw a line horizontally until it meets the two-inch thickness line. Then continue the line straight down, as the colored line shows. The quantities of materials it crosses below the

chart are those you should order, with a ten per cent allowance for waste included. The total concrete required in the example is 2.85 cubic yards. If you were buying ready-mix, this is how much you'd order, upping it to an even three cubic yards.

To mix it yourself, the quantities depend on whether you use the 1:2:3 Strong Mix or the 1:2-1/2:4 Economy Mix. For the Strong Mix the chart shows you'd need 15 ½ bags of portland cement, 1.4 cubic yards of damp sand, and a little more than two cubic yards of stones. For the Economy Mix the chart shows you'd need almost 15 bags of portland cement, about 1.6 cubic yards of sand and 2.3 cubic yards of stones.

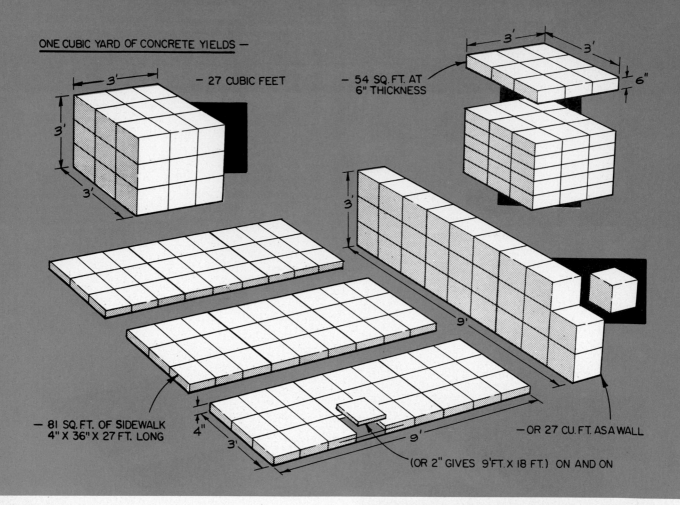

ONE CUBIC YARD OF CONCRETE YIELDS —

— 27 CUBIC FEET

— 54 SQ. FT. AT 6" THICKNESS

— 81 SQ. FT. OF SIDEWALK 4" X 36" X 27 FT. LONG

— OR 27 CU. FT. AS A WALL

(OR 2" GIVES 9'FT. X 18 FT.) ON AND ON

METRICATION

The world is coming rapidly to the metric system. Which system you employ makes little difference to the job, but until everyone uses the metric system and all products come in metric measure, it will help to be able to go from one system to the other. Here's a table of approximations that will help you make the transition most easily:

USEFUL METRIC APPROXIMATIONS

10 millimeters (mm)	3/8 inch
25mm	1 inch
100mm	4 inches
300mm	1 foot
1 meter (1000mm)	3 feet 4 inches
1 square meter (m²)	10¾ square feet or 1.2 square yard
1 cubic meter (m³)	35 cubic feet or 1.3 cubic yard
1 liter	1.06 quart
4¼ liters	1 gallon
1 kilogram (kg)	2.2 pound
45½ kg	100 pounds

Length and width of the area to be covered with concrete, plus slab thickness, governs how much is needed. Here Stanley 25' Powerlock rule is used for the length and width measurements.

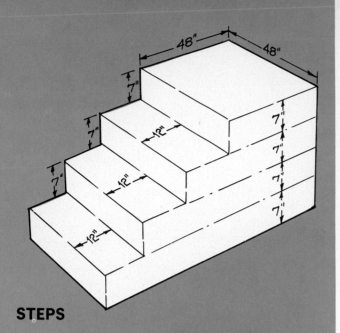

STEPS

Steps might seem tough to figure, but they aren't. If you'll consider them as separate stacked-up slabs, they calculate easily. In the drawing the steps are composed of four 7-inch-thick slabs, each one 48 inches wide. The top-most slab measures 7 x 48 x 48 inches. The next one down is 60 inches long and the next 72 inches long. The bottom one is 84 inches long. Your figuring would go like this:

$$7 \times 48 \times 48 = 16,128$$
$$7 \times 48 \times 60 = 20,160$$
$$7 \times 48 \times 72 = 24,192$$
$$7 \times 48 \times 84 = 28,224$$
$$88,704 = 1.9 \text{ cu. yd.}$$

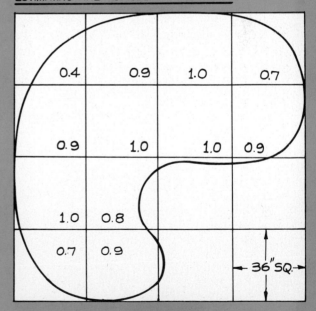

TYPICAL FORMS FOR SIDEWALKS

2" x 2" STAKES

CUT THESE TO CLEAR STRIKE BOARD

2" x 4" FORMS   4" MAX.

BRACES CAN BE EXTENDED OUT TO FIRMER GROUND IF NECESSARY

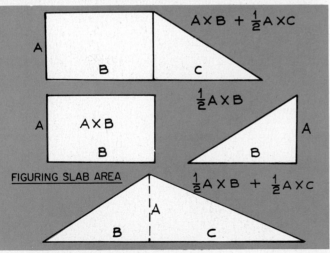

$A \times B + \frac{1}{2} A \times C$

A   B   C

$\frac{1}{2} A \times B$

A   A×B   A
      B

B

$\frac{1}{2} A \times B + \frac{1}{2} A \times C$

A
B   C

FIGURING SLAB AREA

FIGURING OTHER WORK

Calculating how much concrete to order for a wall is little different. A wall is simply a vertical slab. Multiply its length times its height, both in feet, divide by nine to get square yards, then multiply by its thickness in inches and divide by 36. Or look on the back-cover graph.

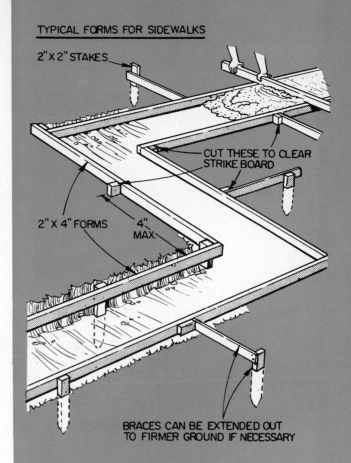

ESTIMATING AREA OF ODD SHAPED SLABS

0.4	0.9	1.0	0.7
0.9	1.0	1.0	0.9
1.0	0.8		
0.7	0.9		

←— 36" SQ.—→

— DRAW TO SCALE ON SQUARED PAPER

— COUNT THE WHOLE SQUARES (4 IN ABOVE)

— ADD ON THE PARTIAL SQUARES BY TENTHS (4.4 ABOVE)

4 + 4.4 = 8.4 SQ. YARDS
OR ALMOST ONE CUBIC YARD AT 4" THICK

Using Sacked Concrete

Sacked concrete makes up for its extra cost in time saved. But its best use is for small quantities.

Prepackaged, sacked mixes come three ways: gravel-mix, which is ordinary concrete containing stones; sand-mix, containing no stones; and mortar mix which is used only for masonry jobs.

THE PRICE OF PREPACKAGED sacked concrete—the kind where all you do is add water and mix—makes it cost almost twice that of ready-mix, but don't let it scare you. It may well be the best way to handle your project. It gets you out of buying, hauling, storing, and handling the separate ingredients for mix-it-yourself concrete. Sacked concrete mixes can be used for any one-of-a-kind small projects. Things like casting a small slab for trash-burning, making a few stepping-stones, or setting a line of fence posts solidly. Think of sacked concrete as instant concrete whenever and wherever you need it. It is also practical for other projects, up to a cubic yard, if the saving in effort is worth the extra cost.

GRAVEL AND SAND MIX

Sacked concrete comes in two types: gravel-mix and sand-mix. These names aren't always used, so know what they are. Gravel-mix is a normal concrete mix containing cement, sand, and stones. Use it for casting projects with sections of 1½ inches and more. Sand-mix contains only cement and sand, making it most useful for thin-section projects and for concrete resurfacing. Because it has no stones, it costs somewhat more per cubic foot than gravel mix. It is sometimes called *topping*.

You'll find many brands of sacked concrete; all those we've seen produce excellent concrete that works well and hardens as well or better than that you'd make yourself.

Premixed concrete comes in various sizes with the most common being 45-pound and 90-pound bags. As in most everything, the larger ones are most economical per pound.

Directions for mixing are given on each package. Ingredients are predried, so the exact water content called for on the package works every time. If, to get easier working, you add more water than is specified, you are weakening the mix. On the other hand, if you use less water, you can make a stronger mix.

Because prepackaged concrete is most often hand-mixed, the manufacturers do not incorporate an air-entraining agent so if you're using it in an exposed location, you should add an agent as described in the chapter on mixing your own concrete. If *Darex* is the agent you get, use one teaspoon per large bag of mix.

The right way to use ready-packaged concrete mix, whether you're mixing a whole bag or not, is to dump the sack and mix everything dry. Then put back what you won't use and save it.

Add water and mix, using the proportion of water called for on the sack. Water will not run away if you add a little at a time, mixing it in before adding more. Use a hoe, square shovel.

When all the water had been added, the mix will look like this. Troweling should produce a smooth surface without tearing (too dry) or running (too wet). All you need is, little practice.

An excellent project for sacked sand-mix is an outdoor table made by embedding wrought iron legs in the fresh slab. Hardware cloth, ½" mesh or chicken wire, reinforces center of slab.

A garden pool, if not too large, is another good project for sacked concrete.

1— DIG OUT EARTH SHAPE

2— ADD CONCRETE AND TROWEL THIS UP SLOPE TO FORM SURFACE

ROUND OFF TOP EDGES

LAY 6"x 6" REINFORCING MESH INTO CENTER THEN TROWEL MORE CONCRETE OVER THIS

Curing can be handled in many ways, but one of the easiest methods, if you have a garden sprayer, is to use white-pigmented membrane curing compound and spray it on. You can start soon as the slab can be walked on without leaving marks. The pigment helps you to see what areas you've covered, yet weathers away after the curing period is over. It's available from most concrete supplies dealers.

The Importance of Curing

The maximum strength of concrete can only be brought out by proper curing methods.

CURING IS STRESSED throughout this book because it is so important. It has to do with the love life of the cement gel, the tiny thing that makes concrete harden. The more gels that take place in the hydration reaction, the harder the concrete will get. And the more durable and watertight its surface will become.

Water is the kicker that makes cement gels do their thing, so you want them exposed to lots of it. For a long time, too. That's why the very hardest concrete is cured under water. Yes, concrete will set under water even better than it does above the ground. The next best thing is to keep water present within the setting concrete for as long as practical.

Why not, you ask, just make concrete with a lot more water in the first place? The trouble with that is it won't stay in the mix long enough to do any good. Just long enough to weaken it. The water bleeds out and evaporates, or migrates to the sub-base. As it does, it leaves tiny capillary pores all through the mix. Later these form places for water to freeze and crack away the concrete surface.

So the answer to quality concrete is to keep the plastic mix as dry as you can work with it, but let it *wet* cure.

As we've said, the best way to do this under water. Since your driveway or patio cannot be built under water, some other means will have to be devised to keep the concrete wet. The accompanying photos show the most practical methods for around-the-house use.

Tests have been made of the hardening of concrete. At first it proceeds at a rapid rate, from the initial set that lets you get onto a slab and finish it, to slabs more than 50 years old. The graph illustrates what happens. Hardness gains fast for the first week or so, then slows. But concrete really never stops getting harder. Very old concrete—properly made—is the hardest of all.

The tests show another interesting fact: stop the curing and let concrete dry out, and its rapid strength gain stops, too. The loss in strength can never be recovered, even by more curing. This is why it is important to cure every new concrete project you build long enough to let it gain most of its strength. The practical limit is usually about a week. (We often say six days, curing). Strip after six days and your project will be strong and durable.

Soaker hose, turned face up, makes a fine concrete curer. Simply snake it around your project, hook it up and turn on the water. Be sure to turn it on often enough to keep surface wet.

Wet-sand cure is an oldie, but a goodie. Spread it out either wet or dry, then soak it well with water from a garden hose. Except in the driest weather, it should not need any further wetting.

Straw-cure is effective, too, if you sprinkle it occasionally · to keep the straw always wet. Spread it out thick enough to hide the concrete beneath soon as you can safely walk on the slab.

Roll out polyethylene plastic sheeting over the newly finished concrete slab and leave it there for a week to effect complete curing. You'll note, on unrolling, the poly is wet underneath.

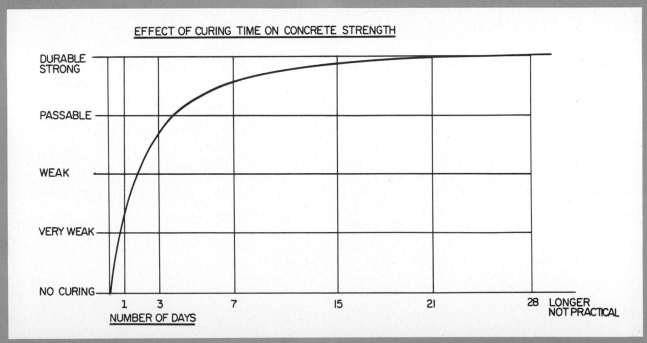

EFFECT OF CURING TIME ON CONCRETE STRENGTH

DURABLE
STRONG

PASSABLE

WEAK

VERY WEAK

NO CURING

1   3   7   15   21   28   LONGER
                                                 NOT PRACTICAL
NUMBER OF DAYS

All About Footings

The chief requirement for a footing is strength, but there are other considerations too. Here is what you should know.

Big footing-casting jobs are ideal for ready-mix. If the truck mixer must be backed up to an open house foundation, don't let it get too close. If the embankment gives way, you'll have trouble.

A FOOTING (or footer) is a below-ground slab of concrete designed to support a wall, column, or other structure. It must be large enough to do the job without wasting concrete; it need not be pretty. Often the excavation it is cast in serves as the form. The top of a footing is merely struck off, perhaps floated once, but left otherwise unfinished. Strength is a requirement; neatness is not. Some footings contain reinforcing steel. The steel need not be there unless the footing is an engineered one. If the steel is needed, then the engineering also is needed. So you may as well omit steel from footings you build.

FOOTING SIZE

All footings serve the same purpose: to distribute concentrated loads from above to as much ground as is necessary to support it. Thus the poorer the ground in bearing capacity, the larger the footing should be to distribute the load. Any footing built on porous or sandy soil should be made extra large. Your building de-partment may have standard footing designs that have been found adequate for your soils. Check with them. If not, ask what size footing they recommend.

For a wall, the footing is often built twice as wide as the wall it is to support. An eight-inch-thick wall thus gets a footing 16 inches wide.

Columns and piers, such as those beneath a porch or deck, concentrate lots of weight on their footings. They carry all of the floor load and all of the roof load for an area half-way to the nearest column on all sides. If a wall is located in this area, they carry its weight too (see the drawing). They also carry what engineers call a *live load* as well as the dead load of the structure. The live load on a floor is furniture and people. The live load on a roof is wind load plus snow. Normal floor live loads are 40 pounds per square foot (psf). Roof live loads vary, but they run from 20 to 50 psf and more depending on snow conditions. Thus, tremendous loads can be brought to bear on footings. Ordinarily, if you make pier and post footings two feet square, they will be large enough.

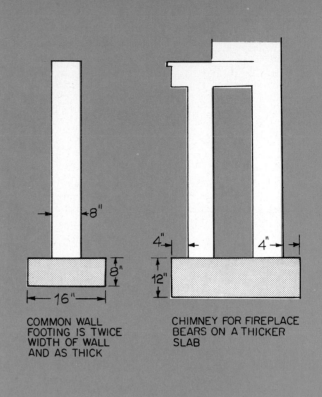

COMMON WALL FOOTING IS TWICE WIDTH OF WALL AND AS THICK

CHIMNEY FOR FIREPLACE BEARS ON A THICKER SLAB

Setting precast concrete piers on their cast-in-place footings in the ground, Anton Ortega provides for concentrated loads from 4 x 4 posts that will rest on them. Keep piers level both ways.

Precast concrete piers come in two sizes— 8″ and 12″—and in several styles: (left), wood block top for toenailing; hole for pegging (center); and (right) nailing strap. Check your dealer.

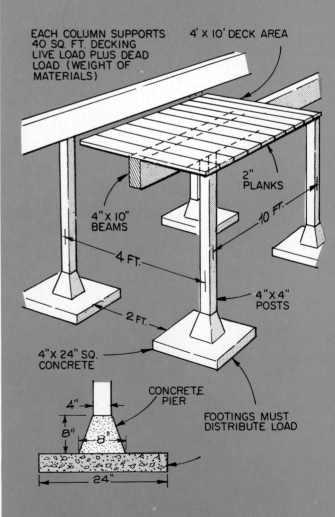

EACH COLUMN SUPPORTS 40 SQ. FT. DECKING LIVE LOAD PLUS DEAD LOAD (WEIGHT OF MATERIALS)

4′ X 10′ DECK AREA

2″ PLANKS

10 FT.

4″ X 10″ BEAMS

4 FT.

4″ X 4″ POSTS

2 FT.

4″ X 24″ SQ. CONCRETE

CONCRETE PIER

FOOTINGS MUST DISTRIBUTE LOAD

Concrete chute is needed to get ready-mix to all parts of a footing pour because truck could reach only two sides directly. Pumping the mix would have worked, too, but was not available.

Footings needn't be pretty or finely finished. Make them strong and just as level as you can, though tie-strips of 1 x 2 lumber hold the tops of the firmly forms together to prevent spreading.

FOOTING DEPTH

Just as important as the size of the footing is depth. The rule applies to all footings—wall, pier, fireplace. Place footings at least 12 inches below frost depth. Build them on unexcavated soil. No footing should be built on fill. No stones are used underneath a footing, even in poorly drained-soil.

Footings are extended up to ground depth with concrete or concrete block walls. Concrete piers or pressure-treated rot-proof wood piers

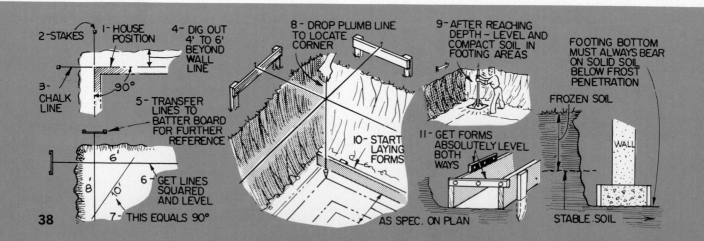

2-STAKES I- HOUSE POSITION 4- DIG OUT 4' TO 6' BEYOND WALL LINE 8- DROP PLUMB LINE TO LOCATE CORNER 9- AFTER REACHING DEPTH - LEVEL AND COMPACT SOIL IN FOOTING AREAS FOOTING BOTTOM MUST ALWAYS BEAR ON SOLID SOIL BELOW FROST PENETRATION

3- CHALK LINE 90° 5- TRANSFER LINES TO BATTER BOARD FOR FURTHER REFERENCE FROZEN SOIL

6- GET LINES SQUARED AND LEVEL 10- START LAYING FORMS 11- GET FORMS ABSOLUTELY LEVEL BOTH WAYS WALL

7- THIS EQUALS 90° AS SPEC. ON PLAN STABLE SOIL

Key, made from an oiled 2 x 4 placed along the center of the footing top after pouring, is pried out with a pickaxe as soon as the concrete hardens sufficiently. Key all block and poured walls.

Footing for a fireplace is formed right along with the wall footings. It should be independent of the house footings, however, since rates of settlement may vary. Pour only on virgin soil.

that can be safely buried below ground are used to bring post and column footings out of the ground. For column footings you can buy precast tapered piers that do the job expediently. Some have steel nail-strips cast in their tops to anchor a wood post to the pier. To make footings, use the Economical Mix described in the chapter on making good concrete.

Footings for complicated or very heavy structures should be designed by a professional engineer and be built to his specifications. The same applies to earthquake-prone areas. Such footings will probably contain steel reinforcing rods or bars embedded in the concrete. If they do, be sure the concrete is well compacted around the steel. Ultimate strength of the job depends on a good bond between the concrete and the steel. Steel reinforcing bars should have at least two inches of concrete cover to prevent corrosion.

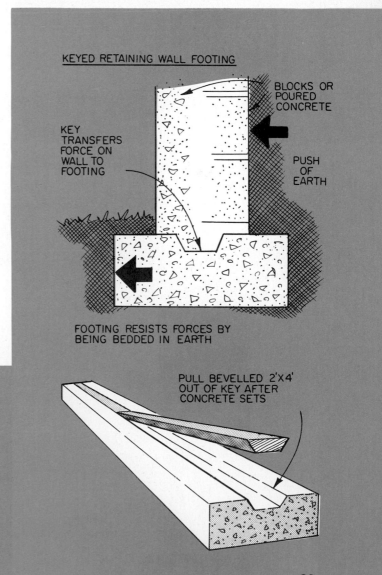

KEYED RETAINING WALL FOOTING

BLOCKS OR POURED CONCRETE

KEY TRANSFERS FORCE ON WALL TO FOOTING

PUSH OF EARTH

FOOTING RESISTS FORCES BY BEING BEDDED IN EARTH

PULL BEVELLED 2'X4' OUT OF KEY AFTER CONCRETE SETS

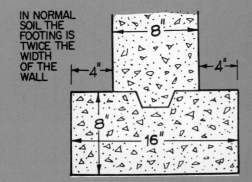

IN NORMAL SOIL THE FOOTING IS TWICE THE WIDTH OF THE WALL

8"
4" 4"
8"
16"

Plain and Fancy Walks

A walk is simple to pour, but it need not be plain and drab. Here are a few techniques for the novice and experienced alike.

A SIDEWALK IS AN IDEAL first project in concrete work. It is usually narrow so you can finish without stepping or kneeling on it. It doesn't take much concrete and you can stop for the day by installing a bulkhead (see drawing) between the forms wherever convenient. At the next session the bulkhead is removed and the resulting keyed joint holds the two slabs together without separation at the joint.

Every concrete project needs forms to contain the concrete while it sets. Sidewalk forms can be 2x4s staked with 1x2 or 2x2-inch stakes driven into the ground as illustrated. Stakes may be up to four feet apart. Front sidewalks are usually built three to four feet wide but a backyard service walk can be as narrow or as wide as you like.

Control joints that create roughly square slabs can be formed by tooling or with leave-in redwood or cypress forms installed at right angles to the side forms.

DRAINAGE

For proper drainage, a sidewalk should slope ¼ inch per foot. Place one form higher than the other or finish the walk with a crown. If the walk slopes along its entire length, no slope or crown is needed.

Control joints should divide the sidewalk into slabs (no more than 1½ times as long as they are wide) to prevent—or to halt—in-between cracking. A sidewalk slab should be about four inches thick. Pour the concrete on well-drained soil; soft earth and sod should be dug out. In poorly drained soil, excavate deeper and place a tamped layer of stones to within four inches of the tops of the forms.

If you dig out too deep and have to backfill, do it in layers. Each layer of earth should be dampened and tamped down hard or else backfill with tamped stones.

HANDLING FRESH CONCRETE

The following tips on handling concrete during placement apply to any project—sidewalk, driveway, patio, footing or wall.

● When moved from the mixer—or ready-mix truck—to the job site, concrete is in danger of

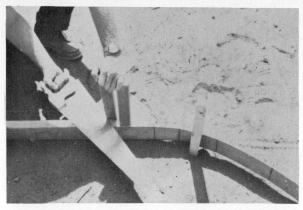

Above. Forming a curve is best done by sawing halfway through the plywood and holding the curve with stakes on the inside and outside. Cut stakes off, remove inside ones while mix is wet.

Technique for avoiding sunken slabs is to provide a good subbase. If slab cannot be cast on unexcavated soil, rake out fill dirt, dampen and compact. Gas powered compacters can be rented.

segregating. That is, its ingredients may drift apart, the stones moving to the bottom of the heap and the water and the cement moving to the top. Segregated concrete is weak concrete, so you should prevent segregation. Do it by transporting concrete in a rubber-tired wheelbarrow, the kind that contractors use. Steel-wheeled wheelbarrows joggle the mix and encourage settling. Air-entrained concrete is resistant to segregation because the tiny air bubbles buoy up the stones and sand particles and hold the water in the mix. Stiff mixes segregate less, simply because they are stiff. It's tough to keep a sloppy mix from segregating, which is another reason why you shouldn't put too much water in concrete.

● You can lay temporary track for your wheelbarrow with planks. These can cross flowerbeds or even go up stairs. Use them wherever needed.

● Don't put too much concrete in the wheelbarrow, it will tend to spill out. Besides, a fully loaded five cubic-foot wheelbarrow is hard to manage. Start small and work up to what you feel you can manage.

● Small amounts of mix can be carried to the

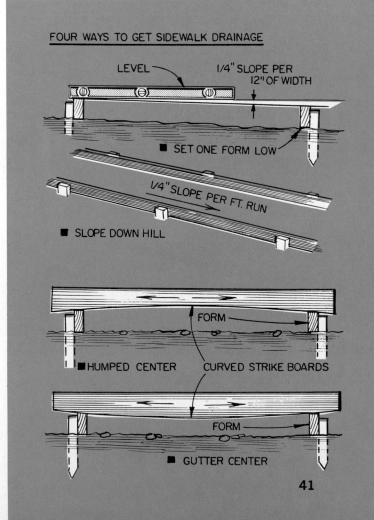

FOUR WAYS TO GET SIDEWALK DRAINAGE

LEVEL — 1/4" SLOPE PER 12" OF WIDTH

■ SET ONE FORM LOW

1/4" SLOPE PER FT. RUN

■ SLOPE DOWN HILL

FORM

■ HUMPED CENTER — CURVED STRIKE BOARDS

FORM

■ GUTTER CENTER

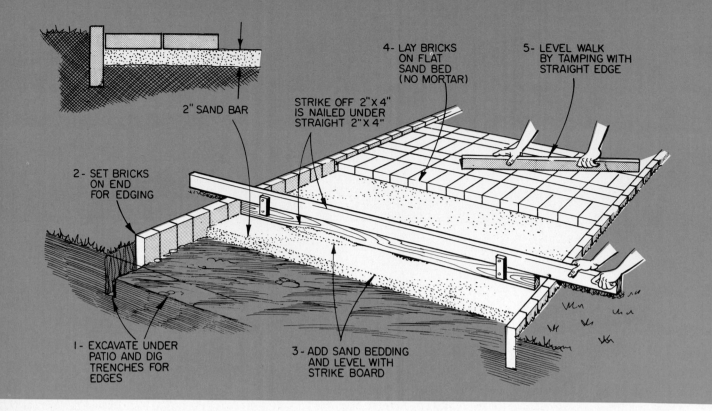

4- LAY BRICKS ON FLAT SAND BED (NO MORTAR)

5- LEVEL WALK BY TAMPING WITH STRAIGHT EDGE

2" SAND BAR

STRIKE OFF 2"x 4" IS NAILED UNDER STRAIGHT 2"x 4"

2- SET BRICKS ON END FOR EDGING

1- EXCAVATE UNDER PATIO AND DIG TRENCHES FOR EDGES

3- ADD SAND BEDDING AND LEVEL WITH STRIKE BOARD

forms in pails, although that's heavy work. If a wheelbarrow cannot be pushed near the pouring site, a bucket brigade is unavoidable. Clean buckets and wheelbarrows after use so that no concrete hardens in them.

DUMPING THE MIX

Dampen the subgrade, with a fine spray, but don't make any puddles or muddy spots. Dump the first load against the forms at one end of the sidewalk. Move it into the corner

with a hoe or shovel and leave it about a half-inch higher than the forms. If you must walk in the fresh mix, wear rubber boots or old ankle-high shoes; concrete in contact with your skin can cause painful burns.

Subsequent loads of concrete should be dumped against the previously placed load. Don't dump them elsewhere and then rake into place.

If you have enough help, someone should start finishing the surface while others bring up more concrete. Strike off, darby, edge, joint, float, and otherwise finish the slab as described

After your forms are filled to the top (a square shovel is ideal for pushing mix) rough screed the surface. Stakes can be left in place although they make the finishing operation more difficult.

After the second screeding you should cut in the control joints and also edge the slab. Use a 2x4 as a straightedge and push down any stones you might hit. After floating touch up the joint again.

42

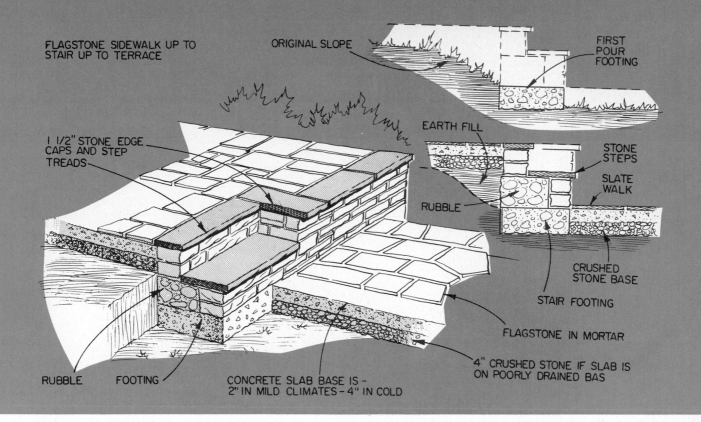

FLAGSTONE SIDEWALK UP TO STAIR UP TO TERRACE

ORIGINAL SLOPE

FIRST POUR FOOTING

1 1/2" STONE EDGE CAPS AND STEP TREADS

EARTH FILL

STONE STEPS

SLATE WALK

RUBBLE

RUBBLE   FOOTING

CONCRETE SLAB BASE IS - 2" IN MILD CLIMATES - 4" IN COLD

CRUSHED STONE BASE

STAIR FOOTING

FLAGSTONE IN MORTAR

4" CRUSHED STONE IF SLAB IS ON POORLY DRAINED BAS

in the chapter on tools and techniques.

SIDEWALK TIPS

Like other slabs, sidewalks should be separated from adjacent walls and existing slabs by *isolation joints*. They're often called expansion joints, though that's a misnomer. The material used is 3/8- or ½-inch-thick asphalt-impregnated joint material. Get it from a concrete supplies dealer; it comes in strips 3½ inches wide and ten feet long. Fasten it to the wall or exist-

ing slab with its top edge about 3/8-inch below the finished level of the slab. During finishing, run the edger along the joint. An isolation joint prevents the new slab from forming a bond with the existing structures and later cracking.

Curved sidewalk forms can be made with one-inch lumber or ¼-inch plywood or hardboard bent to shape. The thinner materials will bend to make sharper curves but stakes are needed both inside and outside the curved sections as well as at the transitions between straight and curved forms. The inside stakes can be pulled as soon as concrete has been

On side walks any expansion joint is used at regular intervals. It may tend to rise during the pouring operation so push it back in place with a straightedged ice chipper or spade.

Most common expansion joint material is asphalt impregnated wood chip material 4"x½" thick. If you can't complete your pour in one day finish at point where there is an expansion joint.

placed around them.

Sidewalks are well adapted to the use of special finishes and colors. You can apply a magnesium float finish or a rubber float finish. Don't do more than one steel-troweling on a sidewalk unless you plan to broom-finish it later because the steel trowel makes the surface too slick for safe use when wet.

Concrete can withstand a light rain shortly after final troweling, but rain during the finishing process or a heavy rain shortly after finishing can spoil the job. The lesson: don't pour when rain is expected!

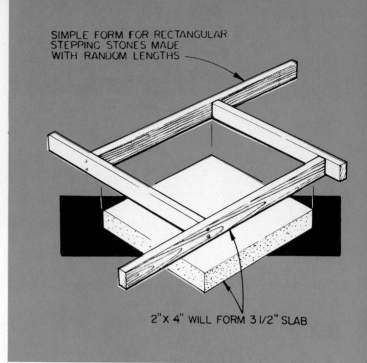

SIMPLE FORM FOR RECTANGULAR STEPPING STONES MADE WITH RANDOM LENGTHS

2" X 4" WILL FORM 3 1/2" SLAB

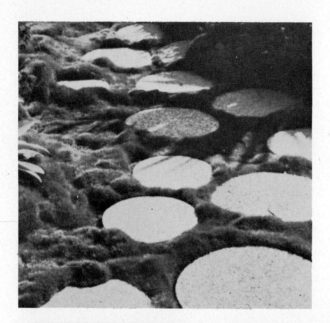

These stepping stones are artistic and are cast in a circular form, finished and cured before hauling to your walk site. They are ideal for hillside slopes where pouring a slab is impractical.

Below. Another interesting and artistic walk treatment. Individuals slabs are first poured and finished. Ordinary fieldstones fill in the spaces and the joints between the stones grouted.

44

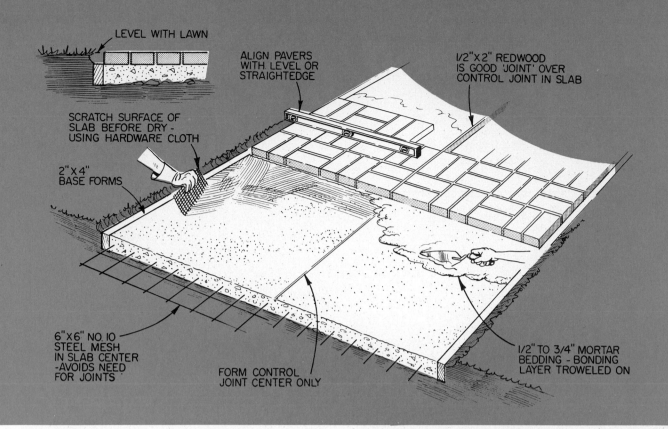

LEVEL WITH LAWN

ALIGN PAVERS WITH LEVEL OR STRAIGHTEDGE

1/2"X 2" REDWOOD IS GOOD 'JOINT' OVER CONTROL JOINT IN SLAB

SCRATCH SURFACE OF SLAB BEFORE DRY - USING HARDWARE CLOTH

2"X 4" BASE FORMS

6"X 6" NO. 10 STEEL MESH IN SLAB CENTER -AVOIDS NEED FOR JOINTS

FORM CONTROL JOINT CENTER ONLY

1/2" TO 3/4" MORTAR BEDDING - BONDING LAYER TROWELED ON

All concrete should cure slowly, and the best way is to keep it damp or wet for the first week. Do this by covering it with plastic, wet rags, damp sod, or with a garden sprinkler. A soaker hose left running slowly will also do the job. Or you can buy a spray-on curing compound such as W.R. Meadows *Seal-Tight Cure-Hard.* Apply it with a pressurized garden sprayer. Curing is important for a strong, durable slab.

HOT AND COLD WEATHER PROBLEMS

In hot weather concrete needs special treatment because it sets up fast. Cut the pour area to a half or one-third. Sprinkle the surface after strikeoff with a fine mist of water to slow air-drying and to cool it. You can start at dawn so as to finish before the heat of the day, or start in the afternoon and finish in the cool of dusk.

In cold weather concrete must be protected from freezing. The best policy is not to start a project if an overnight freeze is possible. Concrete can be ruined by freezing. Even if it doesn't freeze, cold concrete is slow to set.

You can rent kerosene-fired stoves, called salamanders, to prevent your concrete from freezing—if your worst fears are anticipated—during cold weather.

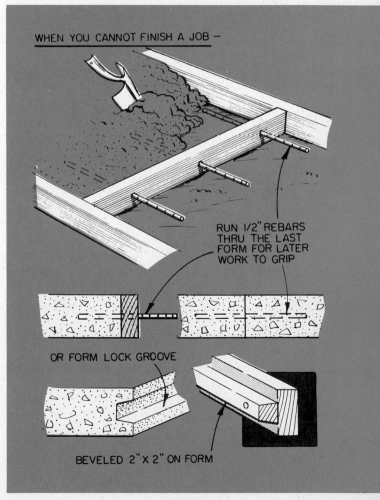

WHEN YOU CANNOT FINISH A JOB —

RUN 1/2" REBARS THRU THE LAST FORM FOR LATER WORK TO GRIP

OR FORM LOCK GROOVE

BEVELED 2"X 2" ON FORM

Pour a Floor

A CONCRETE FLOOR isn't too different from a concrete sidewalk or driveway, but there are differences. An indoor floor need not be made with air-entrained concrete, even in the coldest climate.

Because a concrete floor is surrounded by walls, forming it makes use of what are called *screeds*. Screeds are temporary forms placed around the edges at eight or ten-foot intervals. After strikeoff, the screeds are pulled out and additional concrete is shoveled in to fill the space. This must be done immediately, before the mix starts to set. The fresh mix is blended in with the existing mix by "chunking" it with a shovel. Later darbying smooths it with the rest of the floor. Screeds are staked into the subgrade just as though they were forms. To remove them, wade in with rubber boots and use a crowbar or pickaxe to pull out everything, including the stakes.

Ordinarily, the screeds are placed level all the way around. The level line is established using a garden hose filled with water, or by careful work with a level. Chalk lines snapped onto the walls indicate the tops of the screeds. If you like, the screeds can be nailed to the wall with masonry nails instead of being staked to

the subgrade. Just don't fasten them so well that they cannot be removed easily later.

At any rate, you'll need a strip of isolation joint material between each screed and the wall. Neglect this and the new slab will bond so tightly to the wall it will crack as it shrinks. The isolation joint material should be set 3/8-inch below the level and should extend through locations where the new floor slab meets other slabs, such as where a garage floor meets an existing driveway.

DRAINAGE

Every floor must provide for drainage, to either a floor drain or, in the case of a garage floor, to the driveway. A good many garage floors are sloped to drain towards the driveway. The floor drain, of course, must have somewhere to drain to. This can be a sewer (if codes permit), a dry hole filled with stones, a sump pump pit, a drainage ditch, or whatever other drainage is at hand. A floor drain should not just empty into a stone-filled layer beneath the floor. This encourages rapid settlement that would soon crack the slab. All portions of the floor drain system beneath the floor should be installed before the concrete is poured. Easiest

A floor that will never squeak or rot is a floor that is made out of concrete—and in colors too!

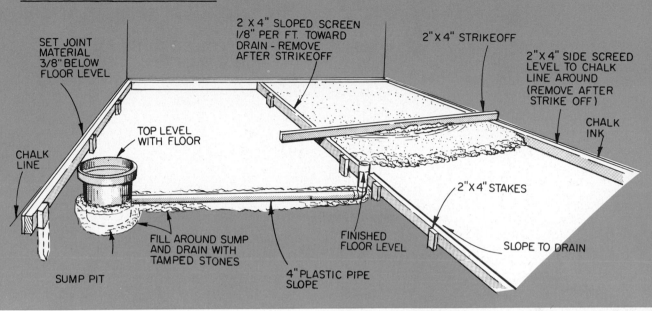

CASTING FLOOR SLOPED TO A DRAIN

SET JOINT MATERIAL 3/8" BELOW FLOOR LEVEL

2 X 4" SLOPED SCREEN 1/8" PER FT. TOWARD DRAIN - REMOVE AFTER STRIKEOFF

2" X 4" STRIKEOFF

2" X 4" SIDE SCREED LEVEL TO CHALK LINE AROUND (REMOVE AFTER STRIKE OFF)

CHALK INK

TOP LEVEL WITH FLOOR

CHALK LINE

2" X 4" STAKES

FILL AROUND SUMP AND DRAIN WITH TAMPED STONES

FINISHED FLOOR LEVEL

SLOPE TO DRAIN

SUMP PIT

4" PLASTIC PIPE SLOPE

to use is four-inch plastic or Orangeburg pipe and fittings. They are widely available at building supply yards.

In areas of heavy rain and poorly drained soil, a basement floor may need inside perimeter subfloor drains to keep water pressure from building up under the floor and cracking it. This is made with perforated pipe placed just inside the forms. Place stones around the pipe to keep concrete out. The drain system is usually run into a sump pit located below the floor, where a sump pump can discharge water into the sewer—if permitted.

An 18-inch diameter concrete or clay sewer tile makes a handy sump pit. Install it with the flange up. You can also buy a plastic sump pit. The top of the sump pit should be at floor level. Run floor drains and the underfloor perimeter drainage pipes into the sump above the halfway point so they'll be above the water level at all times. Failure to do this will cause bubbling and gurgling.

If a basement laundry or shower discharges into the same sump, floor drains should be fitted with traps. Otherwise gases created in the sump will enter the basement. A toilet should never be connected to a sump.

SLOPING A FLOOR

The finished concrete floor should slope

Getting the mix in can be the hardest part of floor-making. In this case a 15' extension trough was rented from the ready-mix producer. Here Jerry Woods nudges concrete with shovel.

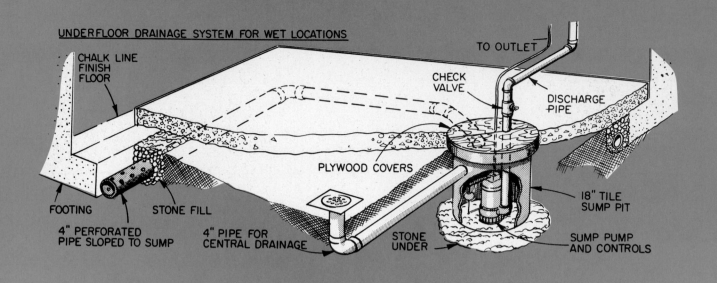

CHALK LINE
FINISH FLOOR

TO OUTLET

CHECK VALVE

DISCHARGE PIPE

PLYWOOD COVERS

18" TILE SUMP PIT

FOOTING

STONE FILL

4" PERFORATED PIPE SLOPED TO SUMP

4" PIPE FOR CENTRAL DRAINAGE

STONE UNDER

SUMP PUMP AND CONTROLS

evenly toward its drain—in the floor or out the garage door. The best way to slope a floor to a more or less central drain is to install additional screeds from the four corners of the walls to the drains (see drawing). Slope these at least 1/8-inch per foot. Like the edge screeds, the sloped screeds are removed after strikeoff. Fill in with extra concrete and chunk it to mix thoroughly with the previously installed concrete. Smooth off with the darby later.

On floors sloped to a floor drain, work the darby with its blade pointed toward the drain. This will maintain the slope, smoothing it without changing it.

Concrete should not come in contact with untreated wood because that encourages rot. Slope of a garage floor should be about an inch for every ten feet, toward the door of course. Get it by sloping the side screeds. In a two-car garage you may need a center screed to keep the strikeoff board shorter than 12 feet. Or you can make what are called "wet screeds" (see photo on page 11).

Floors should be a nominal four inches thick. The subgrade underneath is sloped to maintain this thickness throughout the floor. Any backfilling done beneath the floor should be with compacted stones. Use a four-inch layer in poorly drained soil. Lay a sheet of 2-mil polyethylene sheet over the stones if there's a chance of water coming in contact with the finished floor slab from below. Run it up the walls behind the isolation joints and cut off the excess later.

Ready-mixed concrete may be chuted in

Steel ¼" x 10" anchor bolts are wiggled into the fresh concrete mix and are located 4' to 6' apart around the slab edges of this garage slab. These hold the structure to the foundation.

through a basement window. For a garage floor, the ready-mix can usually be dumped directly from the truck onto the subgrade.

Cut control joints into the floor slab to divide it into ten-foot maximum sections. Basement columns and floor drains are weak spots. Control joints should run to them as shown in the drawing. For a crack-free floor, box the

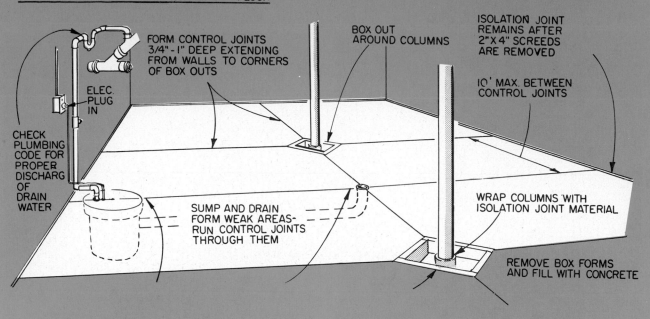

FORM CONTROL JOINTS 3/4"-1" DEEP EXTENDING FROM WALLS TO CORNERS OF BOX OUTS

ELEC. PLUG IN

CHECK PLUMBING CODE FOR PROPER DISCHARG OF DRAIN WATER

BOX OUT AROUND COLUMNS

ISOLATION JOINT REMAINS AFTER 2"X 4" SCREEDS ARE REMOVED

10' MAX. BETWEEN CONTROL JOINTS

SUMP AND DRAIN FORM WEAK AREAS- RUN CONTROL JOINTS THROUGH THEM

WRAP COLUMNS WITH ISOLATION JOINT MATERIAL

REMOVE BOX FORMS AND FILL WITH CONCRETE

Below, floating and finishing a floor slab around the edges can be done best from outside the slab. The out-of-reach center portion of the slab must be done working from knee-boards on the slab.

Forming an integral kick-up for a garage wall is easy using a method devised by professional Carl Roth. Stubby 2 x 2's nailed to the outer form hold inner form. Wall is made on floor, then tilted up.

basement lally columns. Later, remove the box forms and add concrete around the columns, finishing it to match the surrounding floor. Few contractors do it this way, but it will ensure a crack-free floor.

Steel-troweling makes the easiest-to-clean basement floor finish. Be sure to cure the floor for at least six days.

Repairing Walks and Drives

Here's a sidewalk problem that can happen when a tree is planted too close to the walk and the growing roots actually lift the slab. You have two choices; remove the section completely and pour a thinner slab or build a ramp-like slope between the adjacent sections to prevent people from stumbling. Or don't plant trees close to the edge of your sidewalk.

Sooner or later the time will come when a little repair work is called for on concrete projects—here's how to do it.

THREE THINGS can happen to sidewalks and driveways: (1) they settle unevenly; (2) they crack, (3) they scale. If you build yours carefully, you may avoid all three problems. Also, never use de-icing chemicals containing ammonium nitrate and ammonium sulfate, as these chemicals react chemically with concrete, causing it to disintegrate.

If the paving around your house shows signs of deterioration, your best bet is to resurface it. Cast a whole new slab over the damaged one. It can be done.

THIN BONDED RESURFACINGS

There are two ways you can do this—with a thin resurfacing that is bonded to the existing slab; and with a thicker, unbonded resurfacing. Which you use depends on how much of a change in elevation can be tolerated. If the old slab has settled considerably and a raise in grade is appropriate, use the thicker unbonded resurfacing. But if only a slight increase in elevation is possible, go to the harder thin-bonded route.

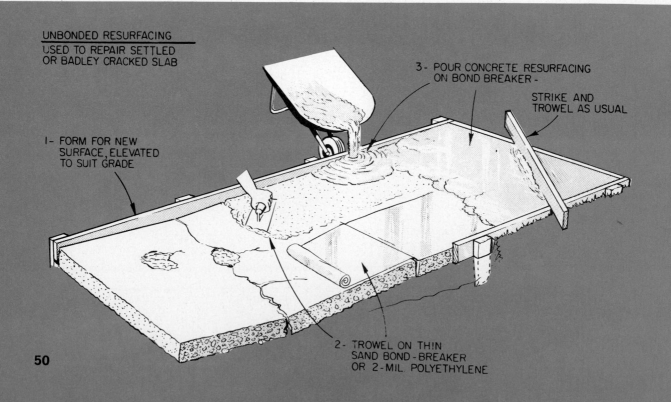

UNBONDED RESURFACING
USED TO REPAIR SETTLED
OR BADLEY CRACKED SLAB

3- POUR CONCRETE RESURFACING ON BOND BREAKER -

STRIKE AND TROWEL AS USUAL

1- FORM FOR NEW SURFACE, ELEVATED TO SUIT GRADE

2- TROWEL ON THIN SAND BOND - BREAKER OR 2-MIL. POLYETHYLENE

To do it, first acid-etch the surface of the old slab with a 20 per cent solution of muriatic acid in water (one part acid to five parts water). This gives "tooth" to the surface for a better bond. Rinse with water until entirely free of acid. Make the form for the bonded topping layer to the desired grade and slope, leaving a minimum of ¾-inch for the topping layer. If the old slab has not settled, you can use construction adhesive to fasten redwood ¾ x ¾-inch strips to the edges as forms. Put additional strips over all cracks in the old slab, because the new topping will crack there, too. (Or else form joints over the old cracks with a jointing tool— if you remember where the old cracks were!)

Mix the topping using 1 part portland cement and 2½ parts of concrete sand. Or you can use sacked topping mix.

Just before pouring it, scrub a slurry of pure portland cement and water into the base slab with a stiff scrubbing brush and place the topping before it dries white. Strike off, edge, finish, and cure as usual.

If portions of the topping are to be *thinner* than ¾-inch, better use a commercial bonding agent or concrete glue instead of the cement slurry. Three brands are Sears *Concrete Patcher*, Coughlan *Permanent Concrete Patch*, and Silcoa *Aqua-Dri Plus*. You can use any one of these.

If the topping layer is to be less than a half-inch thick or feather-edged down to the old slab at any point, the topping mix should be made with a mix-in bonding agent such as *Weld-Crete*, by Larsen Products. Then, no scor-

The best patches are made with prepared patching mixes. Add water and mix according to the directions of the manufacturer, then trowel. For a lasting job, the patch should be wet-cured.

ing or acid-etching is required. *Weld-Crete* also works well as a surface bonding agent.

UNBONDED RESURFACING

Building an unbonded resurfacing is like making a new slab using the old one for a subbase. Form it as usual, making it from 1½ to 6 inches thick depending on the loads it will carry. Before pouring concrete, cover the surface of the old slab with a thin bond-breaker of sand or a sheet of 2-mil polyethylene plastic.

Control joints can be placed anywhere since those in the unbonded base slab will not "telegraph" up to the upper slab. Thin-section slabs less than 3½ inches thick should be jointed into two-foot squares. Otherwise follow the usual jointing requirements. Cure as for any concrete.

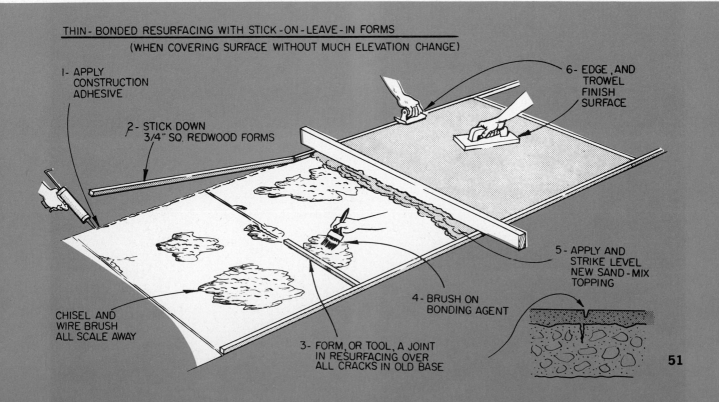

THIN-BONDED RESURFACING WITH STICK-ON-LEAVE-IN FORMS
(WHEN COVERING SURFACE WITHOUT MUCH ELEVATION CHANGE)

1- APPLY CONSTRUCTION ADHESIVE

2- STICK DOWN ¾" SQ. REDWOOD FORMS

6- EDGE, AND TROWEL FINISH SURFACE

5- APPLY AND STRIKE LEVEL NEW SAND-MIX TOPPING

4- BRUSH ON BONDING AGENT

3- FORM, OR TOOL, A JOINT IN RESURFACING OVER ALL CRACKS IN OLD BASE

CHISEL AND WIRE BRUSH ALL SCALE AWAY

51

A concrete driveway is a big do-it-yourself project. But if you have help, and proceed in easy-do stages, you needn't worry. It goes even better with labor-saving tools, like the clever no-stoop straightedge illustrated. Make one from a 10′ 2 x 6 that has a slight crown. Install the 1 x 3 arms so that the crown of the 2 x 6 faces up. It will leave your slab high in the center and eliminate puddles of water that accumulate.

Pour a Driveway

Why pour a plain driveway when with a little imagination —and muscle—it can be made to enhance your home.

DRIVEWAYS are simply wide sidewalks, but because of their extra width, more concrete is required. A driveway usually calls for the use of ready-mix. Few people would care to hand-mix enough concrete for a whole driveway. Though, if you divide the drive into four-foot squares and do one or two squares each day, you bring the job within the reach of your muscles and back.

When most people think of a driveway, they picture a plain concrete slab slightly wider than a car and extending from the street to the garage or carport. With a little imagination, a driveway can be much more. It's out there in front. It may as well add to the appearance of your home. For this reason we recommend that you plan your driveway in a special shape and with a colored, textured, or colored *and* textured finish that will set off the house. Leave-in forms of redwood, cypress, or pressure-treated lumber go well in a driveway.

A good width for a one-car driveway is 10 to 14 feet; make it 14 feet wide if it curves. For two cars parked side by side, the drive should be 16 to 24 feet wide. The drive can be single-width, widening to double-width near the garage and should allow a car to swing into either garage stall. Economical strip drives are satisfactory, if they are straight, but the added maintenance of the center area makes them undesirable.

SLOPE

The maximum slope for an inclined driveway should be a 14 per cent grade which is a rise of 1¾ inches per foot. Changes in grade from sloping to level should be gradual so that cars with long wheelbases don't scrape as they pass.

Good driveway drainage calls for a cross-slope of 1/8 to 1/4 inch per foot, or else a general slope toward the street. A cross-slope can be made by using a curved strikeoff board or by making one side form higher than the other. If the drive slopes to the garage, you'll have to provide a drain at the garage for the water. The driveway should be one inch lower than the garage floor to keep the water out, then flared up to the garage floor in the last six inches of driveway. In fact, it is a good idea to slope your garage floor slightly (1/8 inch to the foot) from rear to front just in case a sudden rainstorm floods your driveway. Also, the slight slope permits cleaning by hosing the floor.

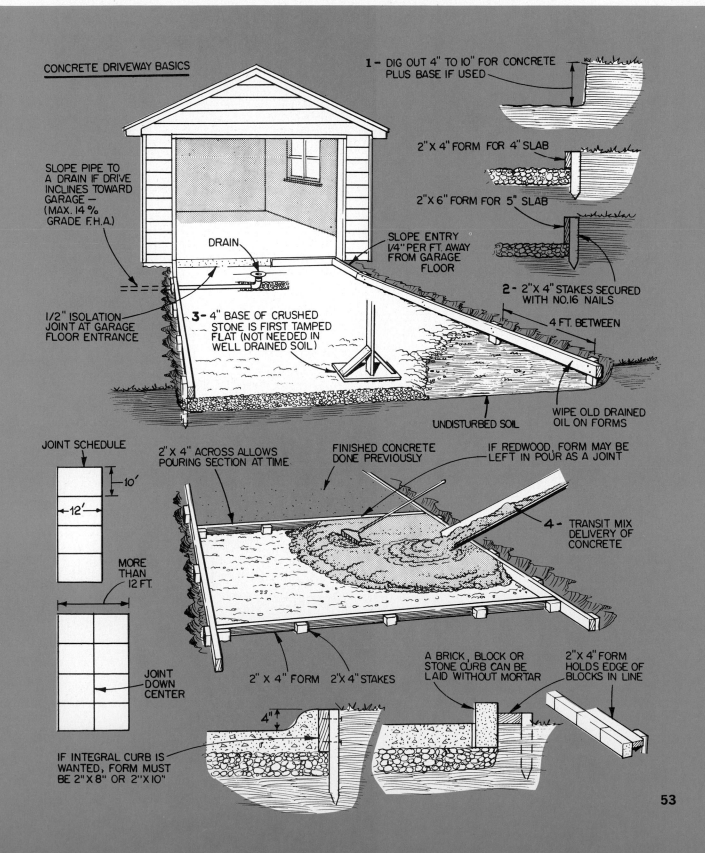

CONCRETE DRIVEWAY BASICS

1 - DIG OUT 4" TO 10" FOR CONCRETE PLUS BASE IF USED

2"x 4" FORM FOR 4" SLAB

2"x 6" FORM FOR 5" SLAB

SLOPE PIPE TO A DRAIN IF DRIVE INCLINES TOWARD GARAGE — (MAX. 14% GRADE F.H.A.)

DRAIN

SLOPE ENTRY 1/4" PER FT. AWAY FROM GARAGE FLOOR

2 - 2"x 4" STAKES SECURED WITH NO.16 NAILS

4 FT. BETWEEN

1/2" ISOLATION JOINT AT GARAGE FLOOR ENTRANCE

3 - 4" BASE OF CRUSHED STONE IS FIRST TAMPED FLAT (NOT NEEDED IN WELL DRAINED SOIL)

UNDISTURBED SOIL

WIPE OLD DRAINED OIL ON FORMS

JOINT SCHEDULE

10'

12'

MORE THAN 12 FT.

JOINT DOWN CENTER

2"x 4" ACROSS ALLOWS POURING SECTION AT TIME.

FINISHED CONCRETE DONE PREVIOUSLY

IF REDWOOD, FORM MAY BE LEFT IN POUR AS A JOINT

4 - TRANSIT MIX DELIVERY OF CONCRETE

2"x 4" FORM   2"x 4" STAKES

A BRICK, BLOCK OR STONE CURB CAN BE LAID WITHOUT MORTAR

2"x 4" FORM HOLDS EDGE OF BLOCKS IN LINE

4"

IF INTEGRAL CURB IS WANTED, FORM MUST BE 2"x 8" OR 2"x10"

Double exposed-aggregate driveway with its ends flared into the street says "welcome" to friends. Exposed-ag concrete is more difficult and time consuming to place and finish than the usual trowel-finished kind. You can do it if you take the time.

HOW THICK?

For cars only, a good concrete driveway need be only be 3½ inches thick, but no less anywhere. If you go this route, you'll have to remember that fuel delivery trucks, garbage trucks, refuse, service trucks, and ready-mix trucks must be kept off. A five-inch thickness is safer and will handle some truck traffic. However, for regular use by heavy trucks—not just pickups—make the driveway thickness six inches. A four-inch thickness calls for excavating half an inch below the bottoms of 2x4 form boards. The five- and six-inch thicknesses require the use of 2x6 forms.

No steel mesh or reinforcing bars are needed in driveways. Steel mesh costs about the same as an inch thickness of concrete. The concrete will do you much more good.

The subgrade should be unexcavated earth or compacted stones. The stones are needed only in poorly drained soils, such as heavy clays in wet climates. Prepare the subgrade, measuring down from a straightedge placed across the tops of the forms. Try for a uniform thickness all over, but don't worry too much about spots that are too deep.

CURBS

You can cast integral curbs on your driveway by using 2x8-inch side forms and striking off with a template made of ¾-inch wood and nailed to the straightedge. The ends of the template are curved to form the curbs you strike off. It can even slope the driveway if the slope is built into the template (see drawing). The embryonic four-inch curb left after strikeoff is finished by floating and troweling along the contour.

Dampen the subgrade just before the ready-mix truck arrives to keep the mix from giving up its water to the subgrade. This should be done with every slab-casting project.

As in the case of pouring a wall, a keyed joint left at the end of a session of work lets you start again at a later time without destroying the slab's continuity.

To get smooth slab edges, tamp the concrete along the forms with a straight-backed shovel or ice chopper.

Be sure to form a cross-wise control (expansion) joint at least every ten feet along the driveway. A driveway more than 12 feet. wide should have another control joint running the full length, usually along the centerline, as shown in the drawing.

Depth of the center joint should be about one inch on a four-inch driveway. You'll also need isolation joints at all adjoining walls and slabs, including the garage floor and sidewalk.

A magnesium float makes a good driveway surface; steel-troweling is too smooth. If you use a steel trowel, finish by brooming to make a nonslip surface. Cure for at least six days before stripping the forms. Then the drive can be opened to traffic.

Local building regulations may govern driveway design. Check with your building department. Normally the driveway meets the street two inches above the street elevation.

Stacked pattern of precast concrete flagstones creates a different-looking driveway. The stones can be cast one or more a day in the backyard, cured, then set on a compact sand subbase. The 1″ joints between each flagstone are filled with mortar and then troweled smooth or tooled.

Cast-in-place concrete rectangles help divide a driveway project into "bite-sized" chunks you can easily handle. It's no strain to do one or two squares a day. Grass grows between them.

All masonry entrance attests do the do-it-yourself skills of the home owner. This driveway is pattern-stamped, colored concrete that looks like bricks but is actually tooled rectangles.

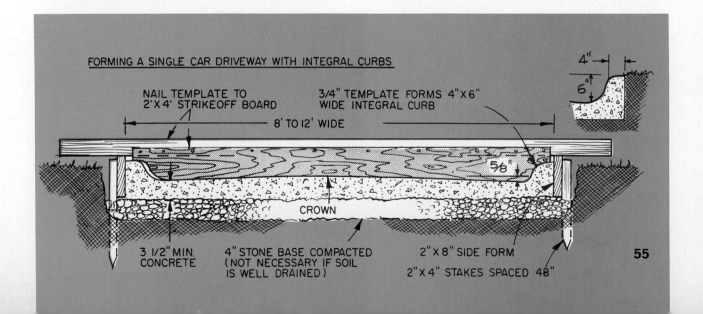

FORMING A SINGLE CAR DRIVEWAY WITH INTEGRAL CURBS

NAIL TEMPLATE TO 2'X 4' STRIKEOFF BOARD

3/4" TEMPLATE FORMS 4"X 6" WIDE INTEGRAL CURB

8' TO 12' WIDE

4"

6"

5/8"

CROWN

3 1/2" MIN. CONCRETE

4" STONE BASE COMPACTED (NOT NECESSARY IF SOIL IS WELL DRAINED)

2" X 8" SIDE FORM

2" X 4" STAKES SPACED 48"

Unusual Surfaces

Concrete need not be drab or strictly utilitarian.
Here are eight different ways to glamorize concrete.

Most popular specialty surface for concrete—and among the most attractive—is exposed aggregate. It's made by brushing away surface mortar to expose the colorful stones beneath.

CONCRETE TAKES EASILY to decorative treatment. In many cases special effects are very little harder to get than an ordinary smooth trowel finish. You need not always use a decorative finish, but you should at least consider one for your project. Very often the special treatment is combined with a colored surface to get both color and texture.

Decorative surfaces can be had by tooling during finishing, by imprinting, by brooming or by otherwise texturing the surface.

EXPOSED AGGREGATE

One of the prettiest ways to enhance surface appearance is to wash and brush away the surface mortar exposing the stones immediately below. This is called an exposed aggregate treatment and it's particularly attractive if special colored aggregates are used in the mix. A wide variety of textures are possible through the use of different-sized aggregates. You can collect your own gravel along streambeds—with the landowner's permission—or buy what you need. A good place to look is at a concrete product dealer's yard or a terrazzo supply house. A broad selection is available. What you use depends on the effect you want. Exposed aggregate is excellent for sidewalks, patios and driveways. It is best used with integral coloring in the concrete.

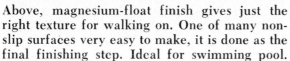

Above, magnesium-float finish gives just the right texture for walking on. One of many non-slip surfaces very easy to make, it is done as the final finishing step. Ideal for swimming pool.

Above right, brooming produces a gritty surface texture on a concrete sidewalk. The degree of texturing may be varied by the time of brooming and broom stiffness. Brush in one direction.

At right, a textured surface was made by sprinkling ordinary rock salt onto the fresh concrete, then troweling it in. Washing out afterwards leaves pock marks, creating the special texture.

Below, a pair of pattern-stamped "brick" projects look like real. Yet both were colored, finished-troweled, then imprinted with a set of rented stamping tools available at most supply houses.

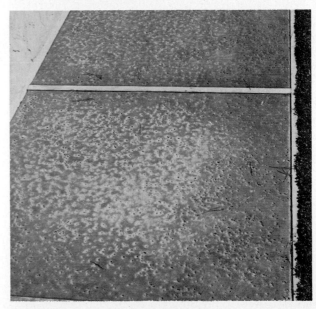

TWO METHODS

You have a choice of making exposed-aggregate concrete with either the mixed-in method or the seeding method. Both are about equal in effort. The seeding method lets you use plain ready-mixed concrete, perhaps colored. The mixed-in method, if special pretty stones are to be used, requires that you mix the concrete yourself.

To seed in exposed aggregate, form and cast the slab in the usual way, then strike it off. That's as far as normal finishing should go. No floating or troweling is required. Let the surface set up slightly then begin spreading the pretty stones over it. Use your hands or a shovel, whichever seems to work best.

Begin tapping the stones into the soft surface. Use a wood float, a darby, or the edge of a 2x4. If the stones sink out of sight, the mix is too soft. Wait until it gets harder. If the stones resist entrance, you have waited too long. Pound them with a brick to force them into the mortar topping. At any rate, the stones should be embedded until they are either flush with the surface or just beneath it. Ideally after embedment, the slab should look as though it had just been floated. In fact, you can run the float over it just to close any openings.

The surface needs time to set up harder before you begin the next step. The rule is to wait until it can hold your weight on knee-boards without leaving much of an impression. Start by brushing away the surface mortar with a stiff nylon bristle push-broom. How much mortar you brush away depends on the size of the stones to be exposed. If they are golf-ball size, you'll have to remove more mortar than if they are pea-gravel size.

FLUSHING

If after brushing you still can't see the stones, just their forms beneath a thin mortar covering, flush the mortar covering away with a fine spray from a hose. This may well be done along with brushing. If the slab has set up hard enough, few stones will be brushed and flushed out.

Brush and flush until the surface has the appearance you want. Don't, however, expose more than the upper third of each stone. If you go deeper, you'll be removing mortar that the stone needs to hold it in place. If brushing and flushing remove mortar too fast or loosen many stones, stop and give the concrete more time to set up.

Treatment too soon while easy work, is

As soon as the concrete has set enough, pretty stones are spread evenly over the surface in a layer one stone-deep. Use a rake or square shovel. The larger stones are best hand-placed.

tough to control. On the other hand, if you wait too long to brush, the surface mortar will be reluctant to come off and hard brushing will be required. In this case, have a wire brush on hand for emergency exposure of problem areas. You'll find that you can expose surface stones quite well when the concrete is hard enough to walk on without knee-boards. If you use leave-in forms, brush the mortar from them with a wire brush and flush all loose mortar from the slab.

BRIGHTENING THE LOOKS

The next day the slab should be washed with a solution of one part muriatic acid (commercial strength) poured into ten parts of water. This removes any dulling mortar film from the surface, brightening the aggregates. Cure the slab in the usual manner.

After curing and air-drying, your exposed-aggregate work will still look dull and lifeless compared to the way it did when wet. To give it a constant wet appearance, treat the surface with a non-yellowing concrete sealer. Those tested and known to produce good results are: *Thoroglaze H* (glossy appearance) and *Thoroglaze* regular (non-glossy), both by Standard Drywall Products, Inc.; Preco *EA Sealer* (glossy); *Horntraz*, by Dewey & Almy Chemical Co (non-glossy); and *Terra-Seal*, by Hilliard Chemical Co. (non-glossy). Apply two coats of the sealer.

Then tap and push the stones down into the surface with a float. All stones should be firmly embedded until they can barely be seen under the surface. Float over them to even surface.

Wait a while, then brush and flush off the hardening surface mortar with hose and stiff broom. Brush deep enough to expose the top third of each stone but do not brush any stones out.

Some dealers carry combination brush-and-flush tools such as this one illustrated. Connected to a garden hose, it does two jobs at once, leaving an exposed-ag surface like that at the right.

59

Using a concrete bonding agent between the courses of an integrally colored two-course job is as simple as brushing it on. The colored second course may be poured immediately or done weeks later. With a bonding agent, topping thickness can be as thin as desired, yet it will still bond well to the base slab.

MIXED-IN METHOD

To make exposed-aggregate concrete by mixing-in, simply substitute the pretty stones for the ordinary concrete aggregates in your regular mix. Handle surface exposure the same as for the seeding method.

To save on the amount of costly aggregates needed and on coloring pigment you can cast a two-course exposed-aggregate slab. The bottom course is a regular mix, even a ready-mix, without special stones. Strike it off an inch below the tops of the forms and finish by scratching the surface with a piece of metal lath or hardware cloth. This step may be omitted if the topping layer is poured right away. Otherwise, apply a concrete bonding agent to the base slab before pouring the top mix containing a coloring and fancy aggregates. This may be days or weeks later, if you wish. The topping need be only as thick as a single layer of the topping stones, but not be thinner than a half-inch. Expansion joints in base and topping should coincide.

Using two-course mix for exposed-aggregate you can get rich-looking effects inex-

pensively because the pretty stones are only surface deep. This system also eliminates most of the mixing, letting you use ready-mix for the bulk of the job, and mixing only the topping yourself.

TOOLING

Tooled and imprinted surface textures are applied during finishing. They may be as simple as lines dividing a slab into flagstone shapes or as complex as some of the imprints illustrated. Simple tooling may be done with masonry jointing tools, the edges of trowels and floats, or even with tin cans. A broom drawn over the surface makes a texture all its own. You can get a good-looking swirl finish by moving the float or trowel in short arcs or circles when finishing the slab. Pattern-stamping is done with aluminum tools, which you can rent.

ROCK SALT FINISH

An arrestingly different finish can be made by throwing rock salt on the surface just after

Colored terrazzo topping mix is spread out over the bonding-agent-coated base slab and struck off. The base course poured the previous day was struck off about 5/8" below top of the form.

Brush-flushing the colored terrazzo topping brings out the beauty of its integrally-cast pea gravel. If brushing tends to knock out many stones, give the terrazzo topping longer to set.

floating and troweling it. Roll or press it into the surface so only the tops of the salt chunks are left exposed. Later, when the concrete has hardened, the rock salt will wash out leaving random impressions. This type of surface is not recommended in freeze-thaw climates because it opens the surface to scaling damage.

Travertine finish is made by spatter-dashing a soupy, tinted topping of cement, water and coloring on the base slab, then troweling it lightly, leaving depressions. Do not use in cold climates.

COMBINATIONS

You can choose combinations of any of the tooled, imprinted exposed-aggregate surfaces, along with others, to get any effect you want. Sometimes a driveway, patio, or walk is made of rectangles of exposed aggregate concrete separated by strips of plain concrete, or by brick strips laid in mortar on a concrete base. Rock salt textures go well with tooling, imprinting, and exposed-aggregate. This reduces the amount of coloring and special aggregates you need to cover an area yet gets away from the too-common smooth patio slab.

If concrete is right for around-the-house work, it's even better when done in color and texture. The next chapter tells how to get the color.

Concrete Can Be Colored

Concrete need not be natural gray or white. It can be colored before, during and after the pour. Any color.

Dust-on method of coloring is economical, putting the color only on the surface of the slab. You can buy the dust-on mix or blend it yourself with cement and sand. Use rubber gloves.

YOU HAVE A CHOICE of four methods for coloring concrete: (1) use an integral coloring agent; (2) apply a dust-on coloring pigment; (3) apply a chemical stain; and (4) paint the concrete. The first two are done during construction. The last two are done afterward, even long afterward.

INTEGRAL COLORING

The best way to color an exposed-aggregate surface or a wall or a cast-in-place or precast item is integrally. For that you'll need to use a coloring pigment. Get only the type made for coloring concrete and mortar. Your concrete supplier should have a selection of colors, including browns, reds, yellow, black, and maybe even green and blue. Popular brands are Sakrete, Pfizer, and Frank D. Davis colors. Cost varies from about 60 cents a pound and up, depending on color. Blue and green cost more than the "earth" colors.

The coloring pigment is added to the portland cement during mixing. (Colored ready-mix can be ordered.) To learn exactly how much pigment to use in creating the color you want, experiment with small samples. Oven-dry the samples before judging them for color tone. Synthetic blacks are potent. About 4 per cent by weight of cement is plenty for a dark color. Other colors need greater concentrations—up to 10 per cent by weight of cement, but never more, because the coloring can weaken the mix.

For rich, clean colors (except blacks and grays) make colored concrete with white portland cement rather than the natural gray kind. It costs more but is best to use for colored work. If you can't find white portland cement, try a terrazzo supplies dealer.

To save on the amount of coloring pigment and white cement, use the two-course method explained in a previous chapter. Be sure to get the amount of coloring pigment in each batch exactly the same. Measure, don't guess. It's best to weight out the coloring on a postal scale.

DUST-ON COLORING

For use with ready-mixed concrete slabs and on large float- and trowel-finished slabs

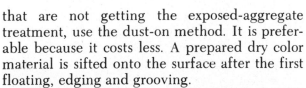

Integrally cast coloring calls for adding color pigment directly to the mix as shown above. For a two-course colored slab, scratch the surface of plain base for a good bond. Add second course.

Concrete stains, as well as oil stains, can be used to color the surface of existing concrete. Vari-colored flagstones tooled into slab get a dark-stain outline. Stains wear very well, apply easily.

that are not getting the exposed-aggregate treatment, use the dust-on method. It is preferable because it costs less. A prepared dry color material is sifted onto the surface after the first floating, edging and grooving.

You can buy a ready-prepared dust-on such as Master Builders *Colorcron* or Dewey & Almy *Colorundum* (containing a nonslip material), or mix up your own. If you mix your own, use a 2:2:1 mixture of white portland cement, fine mortar sand and coloring pigment. You'll need about one pound of dust-on material for each two square feet of surface to be colored.

Apply it on in two applications. After the first dusting, trowel the coloring into the surface with a magnesium float. Dust lightly again and trowel that in, too. Finish the surface as normally and cure without plastic sheeting, which makes a spotty appearance in all colored concrete.

STAINS

Two types of concrete stains are available: solvent-types (like wood stains) and inorganic

types. If the concrete is unpatched and less than a year old, organic stain is recommended because it lasts longer. It changes the color of the surface chemically. A widely used brand is *Kemiko* concrete stain, made by Rohloff & Co. Blacks, browns, beiges, rusts, and greens are to be had.

Solvent stains come in all colors. You can even use those intended for wood, though those specifically made for concrete are best. With any stain, follow the label directions. Application is with brush or roller.

PAINT

The best paint to use for slabs and floors is chlorinated rubber. For heavy traffic indoors and garages use epoxy paint. For walls and floors (except garages), an exterior-type latex will do fine. Portland cement paints may be used on indoor walls, but not floors. With any paint, read the label and follow directions, because some manufacturers use different application techniques. From a cost standpoint, epoxies are the most expensive, Portland cement is least.

Making Mortar

How to prepare a good mortar mix—a most essential ingredient for laying bricks and blocks.

A wheelbarrow, especially one without bolts in the bottom, makes a terrific mortar-mixing box. Dry-blend all the ingredients, first, then add water, and mix with a square shovel or hoe.

MORTAR—the stuff that "sticks" masonry units together—is the weakest part of any masonry wall. Since the wall can be only as strong as its weakest part, you want your mortar to be as strong as you can make it. Unlike ready-mix concrete, you can't buy mortar; you must make it yourself, if only to add water and stir.

Good mortar is plastic and workable in the soft state and it bonds well to the masonry units. Like cement, it hardens after placement. In the hardened state, it should still bond to the units. It must also be tough and weather-resistant. If you start with quality materials and mix them in the correct proportions, you can make good mortar.

BAGGED MORTAR MIX

The easiest way to make mortar for all but the largest masonry projects is to buy a packaged mortar mix. Mortar mix is different from gravel-mix or sand-mix concrete. As with them, all you do is add the required amount of water and stir thoroughly. When preparing only part of a bag of mortar mix, mix the entire bag to reblend the ingredients—and then remove the quantity you need; otherwise you may get a variation because of settling during shipping and storage. A 60-pound sack of mortar mix should be enough to lay 15 eight-inch standard concrete blocks or 30 common bricks with 3/8-inch joints.

MIX-YOUR-OWN

It's less expensive on large projects to make your own mortar starting with the basic ingredients. Mortar is different from concrete in that its workability and bond are just as important as its strength. Therefore, a special cement is made for mortar-mixing, called masonry cement, or *plastic* cement. Neither should ever be used to make concrete. Mix cement with fine mortar sand (also known as beach sand, sand box sand, and by various other names) in these proportions:

> **1 part masonry cement**
> **to 2 to 3 parts damp,**
> **loose mortar sand**

First mix the ingredients dry, then add all the clean drinking water the mix will take without affecting workability. Mortar is the opposite of concrete in this respect. Water is essential for a good bond.

Good mortar spreads easily by trowel, yet doesn't slop off the trowel when picked up. It clings to the faces of masonry units even when they're inverted.

To make colored mortar, dump coloring pigment in with the dry ingredients and thoroughly blend them before adding the water. Use enough of the pigment to get the exact color you want.

Some dealers rent or can sell you plastic treated cardboard or metal mortar-mixing boxes that beat hand-mixing on a concrete floor or plywood base. Mixing is much easier if done with a hoe.

Mortar that stiffens up can be retempered with more water and mixing, but when a batch is more than 2½ hours old (at 80° working temperature), it begins to set and should be discarded. Colder, it will last 3½ hours.

You can mix mortar by hand in a mortar box, wheelbarrow, or on a concrete slab. Do it with a hoe or shovel. However, a better mortar mix can be made easier—and faster—in a powered concrete mixer.

Above, regular cubic foot packaged mortar mix is dumped into 5-gal. mixer cans. Below, one sign of a good mortar mix is that it clings to a plasterer's hawk, even when hawk held upside down.

COLORING

Pigments, the same as described for concrete coloring, may be added during mixing to make colored mortar in yellows, reds, browns, greens, blues, and blacks.

If you prefer, mortar can be mixed with portland cement instead of masonry cement by adding hydrated lime as follows:

1 part portland cement, 1 part hydrated lime, 4 to 6 parts damp, loose mortar sand.

Mix the lime with water first to form a lime putty, then add the other ingredients.

A mortar mix should stand for five minutes after all the water has been added and then mixed again. Or else it should be mixed continuously for five minutes before use.

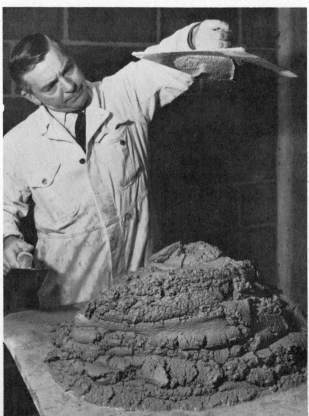

Working With Mortar

Keep your mortar close to hand by using a mortar board—and have an assistant mixing it as you use it up.

Good mortar handling calls for keeping it off the faces of the units. Excess mortar that oozes out of a joint is sliced off with the trowel and returned to the mortar board for reuse.

Left, mortar works easier if you "chunk" or slice it regularly while on the mortar board. Then turn it over and over and throw it back onto the board. Water may be added if necessary.

MORTAR GLUES MASONRY units together, yet is the weakest part of any masonry wall. And since the wall can be only as strong as its weakest part, you want your mortar to be as good as you can possibly make it.

Good mortar in its soft state sticks tightly to masonry units, yet is workable with the trowel. In the hardened state, it bonds well to masonry units and resists weathering. If you start with good materials and proportion them correctly, you can make good mortar.

The very easiest way to make mortar for all but the largest masonry projects is to buy a ready-packaged mortar mix. Called mortar mix on the package, it is very different from gravel-mix and sand-mix. But, as with them, all you do

is add the correct amount of water and mix.

When mixing part of a bag of mortar mix, always mix the whole bag first to reblend the settled-out ingredients.

A 60-pound sack of mortar mix is said to be enough to lay 18 8-inch standard concrete blocks or 40 common-size bricks with 3/8-inch joints. These are ideal figures. Actual ones, I've found, come closer to half that.

MIX YOUR OWN

It's cheaper on larger projects to make your own mortar, starting with the basic ingredients. There are two ways to make mortar—with mortar cement and sand or with Type 1 portland cement, sand and hydrated lime. Here are both formulas:

1 part mortar (or plastic) cement; 2 to 3 parts damp, loose mortar sand.

1 part portland cement; ½ part hydrated lime; 4 ½ parts damp, loose mortar sand.

Mix the lime with water first to form a lime putty, then bring on the other ingredients. The sand should be fine sand, not coarse concrete sand. Use all the water you can get in mortar yet still have it workable. Water is necessary for good bond. Mortar must stand for 5 minutes and be remixed before use. One cubic foot of sand makes about the same amount of mortar.

Good mortar spreads easily by trowel, but doesn't slop off the trowel when picked up. It clings to the faces of masonry units even when they're vertical or inverted.

Mortar that stiffens through drying can be retempered by adding water and mixing. However, when a batch gets more than 2 ½ hours old, it should be discarded.

Coloring pigments, the same as for concrete, can be used to make colored mortar in yellows, reds, browns, greens, blues and blacks.

Five-gallon pail, electric powered, mortar-mixer works while you lay. Add as much water as necessary, still keeping the mortar workable. This is opposite to the way concrete is mixed.

Dump your mix out onto an 18″ square mortar board. Mortar board should be positioned where you can reach it while you lay masonry units and at a convenient height. Move higher as you go up.

Hand-mixed mortar is made on a concrete slab (later washed clean), in a wheelbarrow or mortar box like this one. Mix with a hoe and scoop it out onto the mortarboard. Mortar boxes can be rented.

How to Lay Blocks

Laying concrete blocks, you'll soon learn, is easier and faster than laying bricks or stone.

LAYING CONCRETE BLOCKS is no more difficult than bricklaying. And the wall goes up many times faster because of the large size of the blocks. Concrete blocks are especially good for foundations and large expanses of wall where speed of construction is important. The variety of shapes, sizes, textures and colors is tremendous. Best looking are the sculptured, slump and ground-face units. Ask your block dealer about them.

As with brickwork, concrete blocks may be laid in a variety of bonds. Strongest is the run-

ning bond. Each block overlaps the one below half-way. At corners, the blocks interlace like the fingers of praying hands.

A block wall may be 4, 6, 8, 10, or 12 inches thick, according to the width of the blocks used to make it. For most purposes, though, the eight-inch width is used. Standard of the block industry is the 8x8x16-inch concrete block, which builds an eight-inch-thick wall. Along with 8x8x8-inch half-blocks, this is all you need to build almost anything from a simple wall to a barbecue or a house. The four-

Laying concrete blocks is a leisurely job yet it progresses so fast that it's satisfying to do. Setting the corner units is the only part that requires any skill; the rest is simply laying blocks to a taut stringline. And with a little practice, you can master the trick of laying corner blocks using only a rule, plumb and trowel for the mortar. It's easy enough if you start off level and square. Most important is the bottom course.

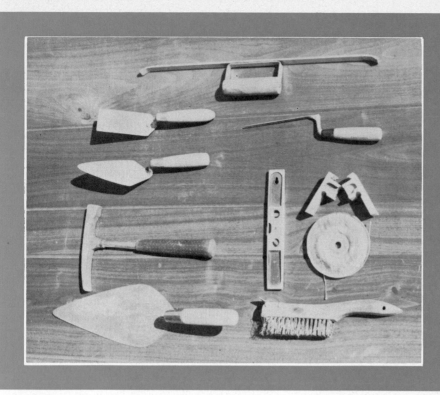

Tools for laying concrete masonry are similar to those for laying bricks and stones. The one special block-laying tool is the 22" long jointer shown at top. Others are: margin trowel, pointing trowel, baby mason's trowel, mason's hammer, level/plumb, line blocks and line, mason's trowel and wire brush.

and six-inch blocks build non-load-bearing walls. The 10- and 12-inch blocks are used chiefly for foundations.

CORE-FILLING

The cores of concrete blocks can be filled after the wall is up. This is done with concrete containing 3/8-inch maximum-size aggregate. Except in earthquake areas or when blocks are used below-ground, the cores usually need not be filled. However, core-filling does make a stronger wall. Thin four-inch block walls make fine low fences. The four-inch-block cores are open only at one end and cannot be filled.

Both two-core and three-core blocks are produced. Two-core ones are more popular in the West. Three-core blocks are more popular in the Midwest and East. Both are made in lightweight and standard-weight. Standard-weight units are made of regular concrete, while lightweight ones are formed with lightweight aggregates such as pumice or cinders. A standard-weight eight-inch block weighs 40 to 50 pounds; a lightweight one, only 25 to 30 pounds.

Blocks with square ends can be used in corners. Otherwise, enough full-end corner blocks must be purchased separately. You can also get jamb blocks, lintel blocks, cap blocks, and bond-beam blocks. Moreover, special units are sometimes available for making chimneys

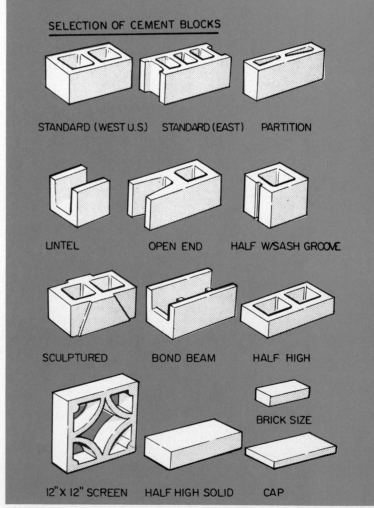

SELECTION OF CEMENT BLOCKS

STANDARD (WEST U.S)   STANDARD (EAST)   PARTITION

LINTEL   OPEN END   HALF W/SASH GROOVE

SCULPTURED   BOND BEAM   HALF HIGH

12"X 12" SCREEN   HALF HIGH SOLID   CAP

BRICK SIZE

and for building pilasters—thick columns in a wall.

Useful for "stretching" a wall is the concrete brick. It's made of the same stuff as a concrete block but in brick size.

Glamor blocks, such as grille blocks, split blocks, slump blocks, can be used to build a concrete block wall that matches brick for beauty.

PLANNING

Because concrete blocks are modules of eight inches—anything you lay up with them should also be modular. Otherwise you'll find yourself cutting and piecing all the way up the wall. To avoid this, see that all your dimensions —height, width, length—can be divided evenly by eight inches. For example: (1) an even number of feet, such as two feet or four feet; (2) an even number of feet plus eight inches, such as two feet eight inches, four feet eight inches; (3) an odd number of feet plus four inches, such as three feet four inches, five feet four inches. These rules apply to vertical as well as horizontal measurements.

Concrete blocks have tops and bottoms. The top of the block has the smoothest, widest face shells (left). Tapered mold makes the bottom face shells narrower. Always lay blocks top side up.

It's easy to estimate how many standard 8x8x16-inch blocks you need for a project. First calculate the square-foot area of the wall less openings. Then multiply by 1-1/9. Thus, a wall of 120 square feet would need 133.3 blocks, a minimum of 134 blocks.

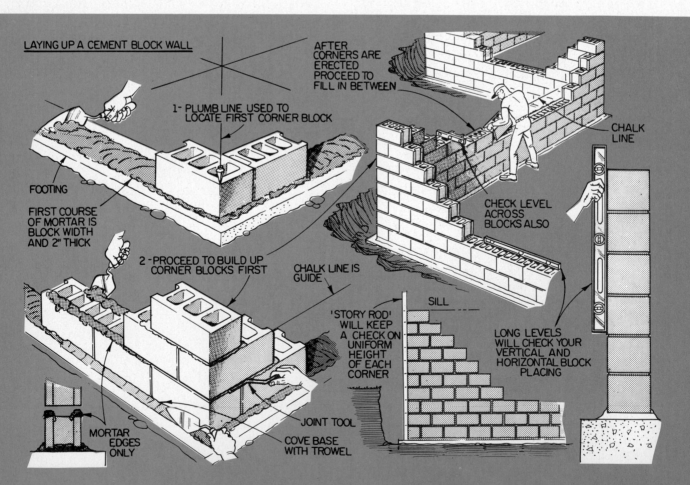

LAYING UP A CEMENT BLOCK WALL

1- PLUMB LINE USED TO LOCATE FIRST CORNER BLOCK

AFTER CORNERS ARE ERECTED PROCEED TO FILL IN BETWEEN

CHALK LINE

FOOTING

FIRST COURSE OF MORTAR IS BLOCK WIDTH AND 2" THICK

CHECK LEVEL ACROSS BLOCKS ALSO

2 -PROCEED TO BUILD UP CORNER BLOCKS FIRST

CHALK LINE IS GUIDE

'STORY ROD' WILL KEEP A CHECK ON UNIFORM HEIGHT OF EACH CORNER

SILL

LONG LEVELS WILL CHECK YOUR VERTICAL AND HORIZONTAL BLOCK PLACING

MORTAR EDGES ONLY

JOINT TOOL

COVE BASE WITH TROWEL

Cutting blocks is quick and easy. First lay the block on a firm surface and tap a line all around, using the chisel end of a mason's hammer. Keep going gently until the block cracks on the line.

Start of a job is to lay out the first course of blocks on the poured footing, dry. Space them with a scrap of 3/8″ plywood and carefully mark the footing (use a chisel) where each joint falls.

Water tube level, with built-in rule at each end is the easy, accurate way to level the corners of a wall. This one, manufactured by Schuyler Products, Kingston, N.Y., fits a regular garden hose.

Story pole with marks for height of each course saves measuring for each course. Set the corner block in mortar and tap it down until its top comes to the mark and is level in both directions.

To lay subsequent blocks from the corner, butter two strips of mortar onto the end of each block. Chunking it down with the trowel helps make the mortar cling tightly and stick to the surface.

Set the end-buttered block down in mortar as you shove it toward the previously laid block in its course. A stringline stretched between corners guides alignment. Level in both directions.

71

Above, to go around a corner with blockwork, butter the ends of a block with mortar and gently push against the side of the already-laid corner block. Tap it level with handle of your trowel.

Below, mortar for the corner is placed so the second-course corner block overlaps the joint in the blocks below. Some masons fill first course. Interlacing of corners units makes for strength.

LAYING THE BLOCK

Start laying blocks as for bricks, on a clean, dampened concrete footing without puddles of water. First lay the blocks dry without mortar to check your modular dimensions and mark where the joints fall. Snap a chalkline between corners. Make a story pole with height marks every eight inches to save measuring corner block heights each time.

Control joints to prevent shrinkage cracking of block walls should be continuous up the wall. For a flexible tie, bridge over them with hardware cloth (½" squares will do) laid in mortar.

TRUE BLOCK DIMENSION

The actual dimension of a block is 3/8-inch less on a side to allow for mortar joints: namely, 7-5/8 x 7-5/8 x 15-5/8 inches.

Unlike bricklaying, in which the corners are first built up to several courses, the easiest do-it-yourself method for laying blocks completes the first course before starting the second. This maximizes the laying of easy in-between blocks to a string line and minimizes the tougher laying of corners. You could do the same with brickwork except that it takes a corner brick some setting time until a string line can be stretched from it without pulling it loose.

With a concrete block you can usually stretch the line almost right after laying the block. So, lay all corner blocks first, right on the footing in a full bedding of mortar. Use a mason's hammer to tap the block into the mortar, but be careful not to chip the face or corner edge where it will show. While it is handier to tap with the trowel handle, because it's already in your hand, this often brings down a rain of mortar on your hand and on the top of the block being laid.

Concrete blocks have tapered cores leaving one surface with a wider shell than the other. (see photo). *Lay blocks with the wide face shell on top.* They're easier to hold this way, and it's easier to spread mortar on the wider shell.

Level corner blocks in three directions: (1) plumb above the block beneath, (2) level lengthwise and at the right height, and (3) level crosswise.

Make sure that bottom and side joints have plenty of mortar before setting a block. No masonry should be joined by stuffing mortar into empty joints. That builds no bond and no strength into the wall. Blocks, after the first course, are laid in what's called *face-shell bedding*. Beads of mortar are placed along the inner and outer face shells of the blocks below. The webs between the face shells get no mortar.

While the mortar is still workable, tool the vertical joints and then the horizontal joints. For horizontal joints the tool used must at least 22" long to get a clean, professional looking joint.

Top off the wall with a course of cap blocks to hide the exposed cores. If additional strength is needed, the cores should be filled. Sometimes a layer of wire mesh is used every 4th or 5th course.

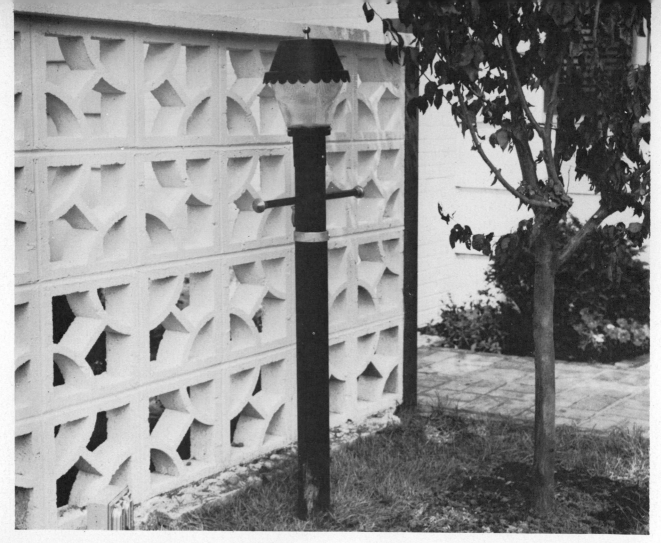

Grille block walls not only give privacy, they break up the wind, letting only a gentle breeze through. This wall was built from 8″ blocks with 12″ solid cap. Walls can be left unpainted.

USE THE RIGHT MORTAR

If the mortar is too wet, it will be impossible to hold the proper 3/8-inch mortar-joint thickness and keep the block course at the right height. For this reason, block-laying mortar needs to be stiffer than brick-laying mortar. If you find your blocks settling too much, add dry ingredients to the mortar in proper proportion to stiffen it up.

Lay the corner blocks for courses so they overlap joints in the ones below. Thus, the corner blocks tie together like the fingers of praying hands.

One difference between blocks and bricks: blocks are never wetted before laying them. In fact, you must keep them from getting wet before and after laying the wall. Between work sessions, cover the pile of blocks with polyethylene sheeting and lay strips of sheeting over the top of the wall to keep rain out of the cores.

Like all masonry work, block walls should

Since grille blocks are cast in economical forms, a multitude of shapes are available. This rectangular unit can be used for decorative walls and also for a chimney by threading over the flue.

74

As you lay grille blocks, match the pattern of each block with those laid previously. The overall pattern of the wall depends on how each block is positioned. Many designs are available.

be protected from freezing until the mortar has fully hardened. Either don't work in freezing weather, or plan on covering and heating the wall while the mortar sets.

You can produce a finished appearance at the top of the wall by laying a course of cap blocks. These have no cores and are two inches thick. Foundation walls need no cap blocks since their tops are covered by the structure above.

Curved block walls can be built (brick, too) by laying the units to a plywood template instead of a line. Curved walls make great fences and are self-bracing against the wind. Pilaster spacing should not be more than twice the height of the wall and the blocks should be tied into the wall either by the bond of the units or by installation of hardware cloth ties in every joint (see photo). Any freestanding wall more than seven feet high, however, should be of engineered design using grouted, reinforced cores.

The trick to laying grille blocks is to make certain they are absolutely straight. Do not lay more than two or three courses a day and use a taut line with each course. Tool flush joints.

After the last course is in place the steel, sill plate and floor joists are added to give the foundation some rigidity. Then the footing drains are installed and the exterior walls that will be below grade waterproofed. Finally, the foundation can be backfilled. Some builders backfill before the joists are in place but this can lead to a collapsed foundation during backfilling.

Building a Dry Foundation

The best time to assure yourself of a dry basement is when you build the foundation.

CONCRETE BLOCK BASEMENTS, heavy soils, and wet weather don't go together. It is almost impossible to keep ground water from going through a concrete block wall and into the basement once it is built. But if you build a basement wall-foundation from scratch, you can built it to be dry. First, let's back up a bit. In well-drained soils or in dry climates, concrete blocks make ideal basements. But with clay soils and in areas with much rainfall, cast-in-place concrete basement foundations are much more suitable. Building them is a job for a contractor. The block basement wall you can build yourself.

In any location a concrete block basement wall should be provided with a watertight exterior coating and an outside perimeter drain system leading down a slope, to a storm sewer, or to a sump pit where the water can be pumped away (see drawing). The footing drains are backfilled with stones. Screen wire is then laid over the stones to keep the earth from the stones. The backfill should slope *away* from the house foundation to carry water away. Downspouts should conduct water away from the foundation.

Foundation coatings of emulsified (water-thinned) asphalt are of little help in waterproof-

ing a block basement wall. Save your money. If the block joints don't crack, then water cannot get through the wall. And if they do crack, the asphalt layer will crack right along with them. Water will infiltrate in spite of the coating. Troweled-on mortar coatings covered with asphalt roofing cement is a much better choice.

BENTONITE PANELS

Bentonite is a fine-grained expansive (not expensive) clay. When water comes in contact with bentonite, it expands to as much as 15 times its original volume. Bentonite panels make use of this expansion to protect foundation walls against water. Bentonite clay is sifted into the corrugations of cardboard panels. These panels are placed over the below-grade portions of the foundation from footing up to ground level. Then the foundation excavation is back-filled with earth. When it rains and water comes against the bentonite, it expands before any water can get through.

Should the block wall develop cracks, the $1/4$ x 16 x 48-inch bentonite-filled panels bridge the cracks and prevent the water from coming through. Even though the cardboard deteriorates, the back-fill holds the bentonite in a continuous sheet over the wall. Not many people—even contractors—know about bentonite panels. To find out where you can get them, write to American Colloid Co., 5100 Suffield Ct., Skokie, Ill. 60077.

You can fasten the panels to the foundation wall with mastic, or nails. Pour a cove of bulk bentonite where the finished foundation meets the footing (bentonite comes in bags). Set the panels into the bentonite cove. Overlap each panel about an inch, as shown in the drawing. Then nail the panels, using concrete nails, along the top edge. Nail through lengths of wood lath, letting the lath extend up above the upper ends of the panels. It thus forms a receptacle for the panels above, and avoids nailing at the lower edges.

A vertical bottom row of panels four feet high and one horizontally placed 16-inch-high row of panels is usually enough to cover the footing up to grade level. Don't cut off any panels until after back-filling. Then you can cut them off and leave the exposed edge to rot away. Or you can leave them sticking up. The edges will soon fall down, adding additional protection at grade.

Bentonite is noncorrosive, harmless to plants and animals and once installed, ruptures in it are self-healing. It comes 15 panels to a bundle, which covers 80 square feet.

INSIDE DRAIN SYSTEM

Bentonite will keep water away from the basement walls, but in really poor-draining soil with a high water table you may need an inside foundation drain. It keeps water from entering through the floor. Use four-inch plastic or Orangeburg pipes sloped toward a sump pit. Most sumps aren't deep enough to work properly. The pit should be dug 42 inches below the basement floor level. Spread a six-inch layer of

WATERPROOFING A NEW BASEMENT OUTER WALL WITH BENTONITE PANELS

FILL BLOCK CORES WITH CEMENT GROUT

BENTONITE PANELS ARE 1/4" X 16" X 48" WITH CORRUGATED CENTER HOLDING BENTONITE FILLER

3/8" X 1 1/2" WOOD LATH BATTEN

PLACE UPPER PANELS BEHIND LATH TO AVOID NAILS

POUR COVE ALONG FOOTING FROM BAG OF BENTONITE

INSERT PANEL BOTTOMS INTO COVE

PANEL MAY BE BENT AROUND OUTSIDE CORNERS

CONCRETE NAILS INTO TOP OF PANELS ONLY

OVERLAP PANELS 1" AT ALL JOINTS

¾-inch stones on the bottom of the pit. Then install an 18- or 24-inch diameter concrete or vitrified clay sewer tile. Back-fill around the tile with more stones. Chisel openings in the tile for the drain pipes below floor level and back-fill around the drain pipes with stones. Install the sump pump and adjust its float switch to maintain the water level at least two inches below the bottom of the lowest-entering drain pipe. Sumps installed too shallow never completely drain the tiles. Sludge deposits build up in them and eventually plug them.

A layer of stone should be placed on the subgrade before casting the basement floor. A one-inch layer of stones over the footing, as shown in the drawing, forms a channel for any water that may seep in through the wall to reach the footing drains without flowing out across the basement floor.

Even though you do not tackle your own construction work, these specifications will ensure a dry basement. They apply, also to cast-in-place concrete foundation walls, although these are not as susceptible to leaks.

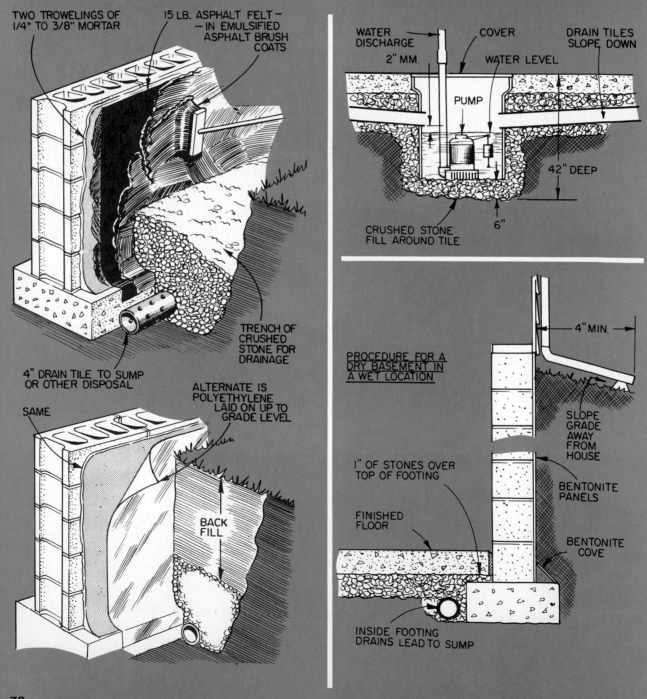

TWO TROWELINGS OF 1/4" TO 3/8" MORTAR

15 LB. ASPHALT FELT — — IN EMULSIFIED ASPHALT BRUSH COATS

TRENCH OF CRUSHED STONE FOR DRAINAGE

4" DRAIN TILE TO SUMP OR OTHER DISPOSAL

SAME

ALTERNATE IS POLYETHYLENE LAID ON UP TO GRADE LEVEL

BACK FILL

WATER DISCHARGE

2" MM.

COVER

WATER LEVEL

DRAIN TILES SLOPE DOWN

PUMP

42" DEEP

6"

CRUSHED STONE FILL AROUND TILE

PROCEDURE FOR A DRY BASEMENT IN A WET LOCATION

4" MIN.

SLOPE GRADE AWAY FROM HOUSE

BENTONITE PANELS

BENTONITE COVE

1" OF STONES OVER TOP OF FOOTING

FINISHED FLOOR

INSIDE FOOTING DRAINS LEAD TO SUMP

If your dozer did not leave you an absolutely level area it is better to dig out the high spots by hand rather than fill in the low spots. Level your footing forms and fill in the sides with dirt.

This foundation was built with 12″ concrete blocks, better (and most costly) than cinder. If block is 25 percent void it is usually classified as solid—used where code required poured walls.

Lay the blocks in face-shell mortar bedding as described in the chapter on block-laying. Lay the corners in each course first, then the in-between blocks. First course is most important.

Corner joints in the footing-drain pipes are made by inserting the pipe loosely into ¼-bend fittings. Split couplings hold alignment while permitting seepage. Keeps basement dry.

Back filling over perforated pitchfiber Orange-burg pipe with crushed stone serves to drain off water that collects there later. It's a necessity in wet clay soil, a good practice with all soils.

Sump pump goes into its pit while the block wall is being built. It keeps water from filling the excavation drain. Polyethylene plastic pipe leads water from site to a dry well, sewer or stream.

If your basement leaks as much as this one you have a real problem that can be solved by patching the inside cracks and looking over the outside terrain to see if rainwater can be detoured.

Waterproof Your Basement

THE BEST WAY to get a waterproof basement is to build it that way in the first place (see pages 76-79) But if you're troubled with a leaky basement that's already built, there are a few things you can do to turn off the "faucet." Ground water usually runs in through a crack in the basement wall or comes in from the joint between the floor and wall. Sometimes it even comes in through a basement window at a window well. Start with the obvious: keep the water from reaching the basement walls. Install four-foot horizontal extensions on all downspouts to lead water away from the house. Be sure that the ground slopes away from the house at least ¼-inch per foot. Cracks between sidewalks, porches, patio, and house wall should be caulked with a non-hardening caulking compound such as a butyl or silicone rubber. If these measures don't work, then *ground water*, not surface water is the problem. You'll have to attack it in other ways.

LEAKING CRACK

Water coming through a crack in a poured concrete basement wall is usually attacked by patching the crack from inside with a fast-set-ting hydraulic cement or epoxy compounds. But it doesn't often work because the wall moves slightly on either side of the crack breaking open any interior patching you do. Even the reinforced epoxy compounds can't hold against such movement. The only sure cure is to seal the crack against leakage from the *outside*. Do it with bentonite, the same earth material described in the chapter on building a dry basement. First dig down next to the crack with a post-hole digger. Use extensions if need be, but dig all the way down to the footing. (If the crack is underneath a concrete slab, you'll either have to remove the slab or call in a professional basement waterproofing contractor who has special equipment to reach in without digging.)

Shove temporary form boards into the hole and arrange them to leave about four inches of space around the crack. Pour the forms full of dry bentonite. One sack should do one crack. Backfill the hole with the soil you dug out, pulling the form boards out as you backfill.

What you've done is build a four-inch bentonite seal around the crack. When it rains, the bentonite will get wet, swell and seal off the crack.

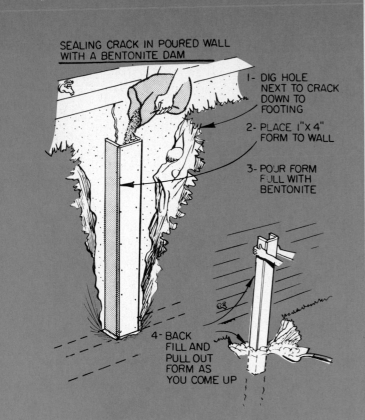

SEALING CRACK IN POURED WALL
WITH A BENTONITE DAM

1- DIG HOLE NEXT TO CRACK DOWN TO FOOTING

2- PLACE 1"x 4" FORM TO WALL

3- POUR FORM FULL WITH BENTONITE

4- BACK FILL AND PULL OUT FORM AS YOU COME UP

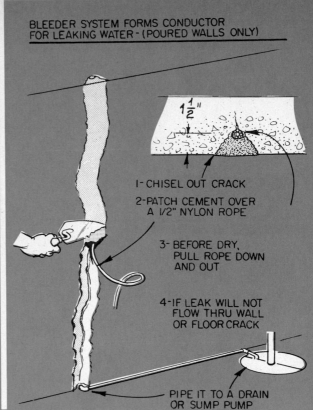

BLEEDER SYSTEM FORMS CONDUCTOR
FOR LEAKING WATER - (POURED WALLS ONLY)

1½"

1- CHISEL OUT CRACK

2- PATCH CEMENT OVER A 1/2" NYLON ROPE

3- BEFORE DRY, PULL ROPE DOWN AND OUT

4- IF LEAK WILL NOT FLOW THRU WALL OR FLOOR CRACK

PIPE IT TO A DRAIN OR SUMP PUMP

Patching leaking basements is usually a trial and error chore. Here are some techniques that work.

In use your bentonite seal may expand and rise as much as a foot above grade. Resist for several months the temptation to shovel it away. After that you can remove the excess and cover it with black dirt. Bentonite will not harm plants or animals.

A bentonite-sealed crack need not be patched from inside unless it's unsightly. To patch the crack, chisel a V-notch along the crack, brush it free of dust and apply a commercial concrete patching compound. If the crack is wide enough for bentonite to sift through from the outside, it will have to be sealed from the inside.

BLEEDER SYSTEM

Cracks that cannot be sealed from outside need what is called a *bleeder system*. The bleeder acts as a hollow tube running down inside the crack and leading water away.

To make a wall bleeder, first chip out the crack 1½ to 2 inches deep in a V shape going from top to bottom. Clean and dampen it. Starting at the top, lay a short length of half-inch nylon rope into the crack and down about a foot, pushing it into the groove of the V. Pack

fast-setting hydraulic cement into the crack and around the rope and trowel it flush with the surface. Then, while you still have enough rope to get hold of, pull the rope down but not out of the patched portion.

Keep patching and pulling until you get the crack patched from top to bottom. Leave an opening at the bottom for the bleeder to drain. It can drain through a hole drilled in the basement floor (provided the floor is built over a stone fill). Or it can be fitted with a pipe or hose that runs to a sump pit or floor drain. You've patched the crack but left a half-inch tube down its entire length to carry away water coming in through the crack from the outside.

BLOCK WALLS

The foregoing applies only to poured concrete walls. Concrete block basements are by far the worst leakers. Sealing them successfully is more difficult. What usually happens is that the walls crack, letting water enter the cores of the blocks. From there it seeps into the basement. You can sometimes even hear water dripping inside the cores!

81

The answer is to drain the cores with a bleeder system designed for this purpose. To make it drill through the shell of the block and into the water-filled core. The water will rush out. Next chip a small drain channel in the floor next to the wall for about four inches on each side of the drain hole. If this collects the drain water and lets it seep through the floor into the gravel layer and sump pit, you're home free. If not, chip a groove in the floor leading to a hole drilled through the floor out beyond the footing. About nine inches out from the wall should do it. Lay a half-inch copper tube in the groove and fill over it with mortar troweled flush with the floor. The pipe connects the core hole and the floor hole.

WINDOW WELL

A leaking window well usually can be cured by draining it with a surface bleeder system (see drawing).

Perhaps your problem isn't leakage at all, but condensation. It's sometimes hard to tell. Fasten a small mirror, such as a shaving mirror, to the basement wall. Look at it a day later. If both wall and mirror are wet, proper basement ventilation or a dehumidifier will prevent condensation. If the wall is wet but the mirror is dry, the problem is leakage.

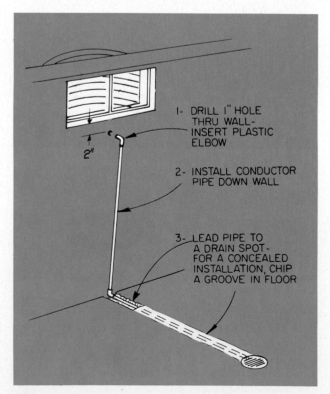

1- DRILL 1" HOLE THRU WALL- INSERT PLASTIC ELBOW

2"

2- INSTALL CONDUCTOR PIPE DOWN WALL

3- LEAD PIPE TO A DRAIN SPOT- FOR A CONCEALED INSTALLATION, CHIP A GROOVE IN FLOOR

On leaky blocks walls (and cracked poured foundations), cracks can be usually made waterproof with hydraulic cement. Chisel out a square or undercut canal ¾″ wide and ¾″ deep.

Make certain you chisel out at least ¾″ between the floor and the wall. Never make a V cut because the water pressure can force the cement out. Rather, undercut joint. Clean out cuttings.

Two part epoxy mix by X-pando forms a tough coating to prevent water penetration. It adheres to any surface and is not affected by water, temperature or weather. Easy to apply.

Chip out horizontal and vertical joints where leaking was noted and flush away dust and cuttings. The joint between the floor and wall is often leaky and should also be chiseled out.

Cracks in wall are filled flush with surface but the base joint is triangular. Push the cement down into the crack, then finish off with a small trowel. To cure keep damp for a least 15 minutes.

Hydraulic cement comes in pint, quart and gallon cans. A pint (1½ pounds) will fill a ¾"x¾" crack 4½ feet long. It sets up fast so mix only small quantities to putty consistency.

A companion product that seals small cracks and pores is mixed with water to brushing consistency. It comes in colors and hides a multitude of sins. Coverage is about 3½ ounces per sq. ft.

X-pando pointing mortar is ideal for repointing brick joints. Unlike regular mortar mix, it actually expands to permanently fill every hairline crack between mortar and the brick.

Another way to prevent moisture penetration is to apply a coat of silicone over the entire surface. Water-clear, it can be brushed or sprayed on. Prevents efflorescence on brick.

Building a Retaining Wall

A retaining wall should live up to its name. Here are a few facts you should know before you plan to build yours.

If your house is much above the street, you can have a level front yard by building a retaining wall along the sidewalk. This brick beauty has been set back for a sidewalk planter bed.

A RETAINING WALL is one of the most risky structures you can build as it can be pushed over by heavy soil pressures from behind. For this reason, most building codes limit retaining wall height to four feet unless the design is engineered. If you must go higher, either consult an engineer or use a series of four-foot-high walls stepped up the hill. Experts maintain you should have engineering advice, too, on slopes greater than 36 per cent. In any case, a building permit is usually required.

To minimize earth pressures on a retaining wall, be sure to provide for drainage so water will not add its pressure to the wall. Wall drainage can be designed in the form of a shaped-earth gutter near the top of the wall and leading beyond it on one or both ends.

Another possibility is to lay a drain tile line along behind the base of the wall and back-filling with a porous stone fill. Sloped to the end of the wall on one or both ends, the line will carry water out before pressures can build.

Through-the-wall drains, in the form of cast-in tiles or holes left in the wall at regular intervals can also serve the same purpose. In any case the water pouring out of them should not be allowed to undermine the wall's footing. Build a paved gutter, if necessary, to catch it and lead it away.

CONCRETE WALL

Strongest, but not the easiest to build, is a massive concrete retaining wall. The drawings show an accepted design built without reinforcing steel. Both footing and wall are cast at the same time. Earth serves as a form for the footing. Plywood, ¾- or 5/8-inch thick, can be used to form the wall, bracing it on 12- to 16-inch centers with 2x4s.

Most of the stress on a set of forms comes from the pressure of the wet concrete pushing outward. This force can be balanced from one wall to another by drilling holes in the form and running tie-wires inside the form between its two walls. Concrete is poured around the tie-wires, which are later cut off. Nails, or sticks can be used to twist the tie-wires tight. Be care-

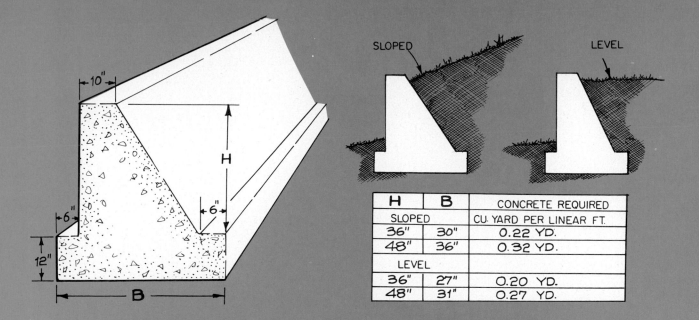

H	B	CONCRETE REQUIRED
SLOPED		CU. YARD PER LINEAR FT.
36"	30"	0.22 YD.
48"	36"	0.32 YD.
LEVEL		
36"	27"	0.20 YD.
48"	31"	0.27 YD.

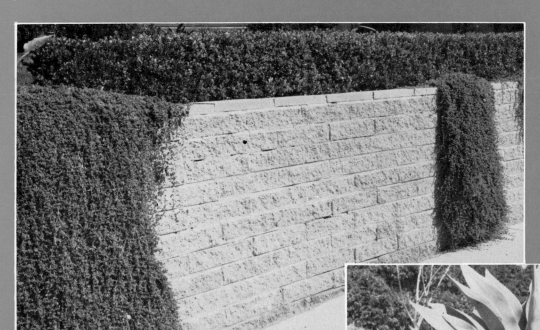

Split concrete blocks make a stonelike retaining wall. Slim split cap blocks cover the hollow cores, which have been filled with grout around reinforcing steel. Poured walls are still strongest.

A retaining wall can be as simple as stacked-up pieces of broken sidewalk. These may be found at some city dumps and sledged into right-sized chunks. Check for nearby concrete demolition.

Low retaining wall forms planter for tree. To lay bricks in a circular pattern, cut a curving template from ¼" plywood or hard board. Stake in place, then lay bricks to top edge. End-wise bricks trim top.

ful not to over-tighten the wires or else you may get a pinched-waist form. Short reinforcing bars, or even sticks, are sometimes used to hold the forms in place as the wires are tightened.

REINFORCED CONCRETE WALL

You can save on concrete and the forms by building an engineered reinforced concrete wall. It can have a much thinner section because the bars take much of the stress put on the concrete. Soil must be firm to prevent tip-over. The drawing shows a design for a reinforced concrete retaining wall up to four feet high. The backfill behind it may be either sloped or level. Install through-the-wall drains six feet apart or closer.

Reinforced concrete must be compacted well around the bars. Tapping on the bars lightly and rapidly with a hammer as you cover them with concrete ensures a firm bond between concrete and the bars.

Where vertical and horizontal reinforcing bars cross each other, tie them together with double loops of heavy wire twisted tight with pliers. You can bend bars with a pipe wrench.

Bars must be continuous or else overlapped by at least a foot and secured with wire.

You can get reinforcing bars from a concrete products supplier or building supply yard. They come in 20-foot lengths. Have them cut or bent in half so you can get them home easily.

OTHER MATERIALS

Retaining walls may also be made of concrete block, brick, and stone. Sometimes low stone walls are laid dry without mortar. In spite of its informal construction, such a wall may last for years. A reinforced brick or block wall may be built to the same dimensions as the one shown for reinforced concrete, but make it of eight-inch-thick concrete block or two tiers of four-inch brick with a half-inch space between them for the bars. The cores of blocks and the between-tier space in a brick retaining wall is filled with grout, a mixture of portland cement and sand, using enough water to give it a pouring consistency. The top of such a wall can be finished with a cast-in-place concrete cap or a course of bricks or cap blocks to hide the grout and reinforcing. See the chapters on bricks, blocks, and stones.

Drainage should be provided at the base of a retaining wall. A neat way to do it is to lay a concrete block sidewise in the wall. Its cores drain the water off. Perforated pipe is also used.

Hammering a through hinge hole into the ground holds this precast on-ground wall in place. Called Wonder Wall, this block is sold by Hazard Products, San Diego, California.

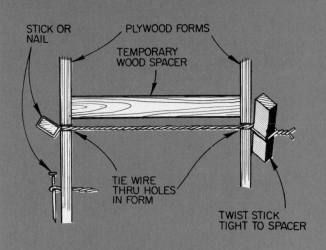

STICK OR NAIL

PLYWOOD FORMS

TEMPORARY WOOD SPACER

TIE WIRE THRU HOLES IN FORM

TWIST STICK TIGHT TO SPACER

FORMING 'CAST-IN-PLACE' RETAINING WALL

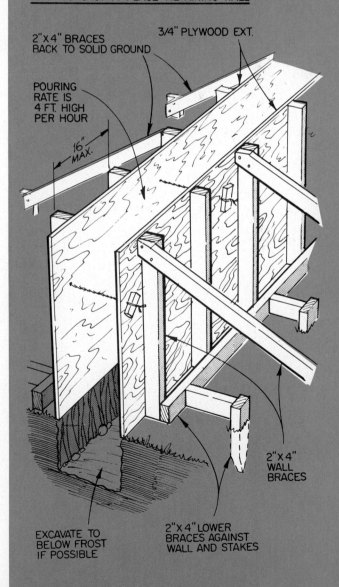

2"x 4" BRACES BACK TO SOLID GROUND

3/4" PLYWOOD EXT.

POURING RATE IS 4 FT. HIGH PER HOUR

16" MAX.

2"x 4" WALL BRACES

EXCAVATE TO BELOW FROST IF POSSIBLE

2"x 4" LOWER BRACES AGAINST WALL AND STAKES

Masonry Fences

Build a fence that never rots, never needs painting, never needs any maintenance and will outlast the house.

ORDINARILY, concrete and masonry are ignored in selecting material for fence-building. Yet they make excellent maintenance-free fences. Most are laid on a footing. See the chapter on footings for rules governing how deep they should be or build a no-footing fence (see drawing). Once the footing has been poured, you can build your fence with brick, blocks, stones, or cast it in place using plywood forms. The easiest to build is a concrete block fence. Special grille block units are available to create a fence that breezes can blow through, yet will tame high winds.

Some concrete block producers make units especially for fence-building. These are four inches thick. Pilaster units twice as thick have recesses that hide the joints between sections of the fence at corners. Check with your dealer to see what's available.

Whatever you use, lay them according to the directions in the chapter covering that kind of masonry.

FENCES AND THE LAW

Any fence you build must meet local laws. These vary from community to community, so find out what regulations will affect your fence before you start planning. Some govern the length, height, and thickness of the fence. Others pertain to setbacks from property lines. Some regulations outlaw fences. Deed restrictions on your title can also govern fences. So can zoning ordinances. Check everything.

Legal problems get even worse if a fence runs along a property line. Sometimes you can build a fence astride the property line without first getting permission of your neighbor. In other cases you can't. Such a fence may become half his even though you pay for it. If you build the fence entirely on your property, you usually needn't get your neighbor's agreement as long as the fence meets all regulations. But you'd better be sure where the property line is—even putting the fence several inches on your side of the line to be clear of any later disputes.

The best bet is to put up a fence that both you and your neighbor agree on. Best yet, get his help in buying materials and construction, and any eventual maintenance.

After the fence is up, you can paint it if you wish. Use exterior latex paint of the desired color. It will give a lasting color on concrete masonry that won't need recoating for a long time.

DURABLE AND DECORATIVE LOW FENCE

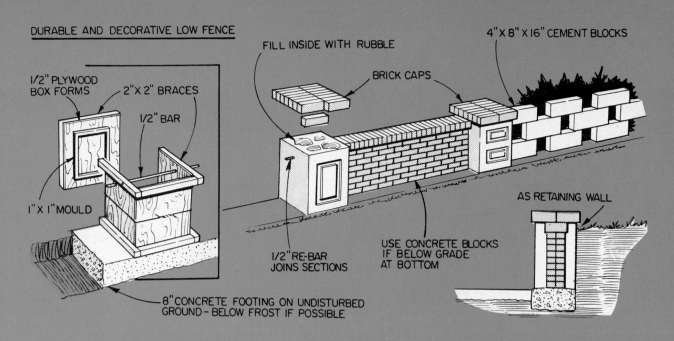

1/2" PLYWOOD BOX FORMS

2"X 2" BRACES

1/2" BAR

1" X 1" MOULD

FILL INSIDE WITH RUBBLE

BRICK CAPS

4"X 8"X 16" CEMENT BLOCKS

AS RETAINING WALL

1/2" RE-BAR JOINS SECTIONS

USE CONCRETE BLOCKS IF BELOW GRADE AT BOTTOM

8" CONCRETE FOOTING ON UNDISTURBED GROUND - BELOW FROST IF POSSIBLE

PATIO OR PRIVACY FENCE FROM CONCRETE SLABS

-BUILT WITHOUT A FOOTING !!

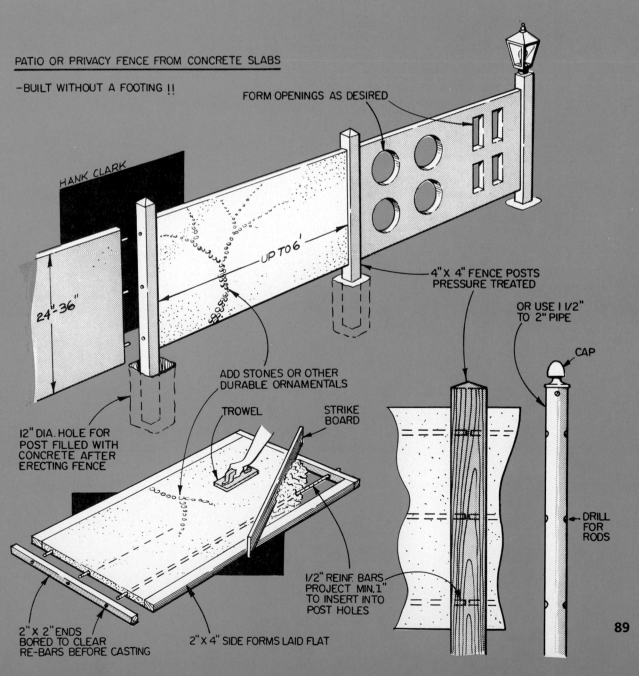

FORM OPENINGS AS DESIRED

HANK CLARK

UP TO 6'

24"-36"

4"X 4" FENCE POSTS PRESSURE TREATED

OR USE 1 1/2" TO 2" PIPE

CAP

12" DIA. HOLE FOR POST FILLED WITH CONCRETE AFTER ERECTING FENCE

ADD STONES OR OTHER DURABLE ORNAMENTALS

TROWEL

STRIKE BOARD

DRILL FOR RODS

1/2" REINF. BARS PROJECT MIN. 1" TO INSERT INTO POST HOLES

2" X 2" ENDS BORED TO CLEAR RE-BARS BEFORE CASTING

2" X 4" SIDE FORMS LAID FLAT

89

How to Lay Bricks

The art of laying brick consists of being patient, careful, and following the accompanying instructions.

ONE OF THE MOST satisfying do-it-yourself projects is bricklaying. It moves along reasonably fast. And with just a little care you can do as good a job as a professional, one that your friends won't believe you did. You cannot work as fast as the pro, that's all. Be satisfied with 50 to 200 bricks a day, not the 1,000 a professional can lay.

Bricks come in all colors and a number of shapes. The best way to pick out what you want is to visit a dealer and see what he has. Some dealers have their bricks displayed with the units laid on strips of fiberboard to resemble their appearance in a wall. This is a big help in selection.

If you're going to lay bricks outdoors, you must either get *medium-weathering* or *severe-weathering* bricks. The high-type units called face bricks are made only in these two types. A third type is called non-weathering, and should be used indoors or in protected locations only.

Bricks in *modular* sizes lay up to make either 8- or 12-inch modules. Non-modular bricks don't come out so even. No harm unless you're

House first-floor wall of 6″-wide SCR bricks here is laid two tiers thick so that the wall can go a second story high. A single story home, built with SCR bricks, may be only one tier thick.

Below right, plastic mason's line blocks make it easy to handle the taut line at wall corners. Move the blocks up as each new course is completed. Line should be flush with the top.

Below. Corners are the most important part of a wall. Take the most care with them, keeping them absolutely plumb, even.

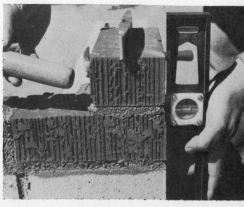

trying to fit in a modular window or door opening. Then you must do some fancy cutting or joint jockeying.

FIGURING QUANTITIES

The chart on page 94 shows how many of each type of bricks are needed to build 100 square feet of wall four inches thick. If your wall is to be a bearing wall, (a wall that supports a floor or a roof) then twice as many bricks will be needed. The brick quantities shown allow a customary 5 per cent for waste.

You can also figure mortar quantities from the table. Because the amount of mortar depends on the thickness of the mortar joints, three thicknesses are shown. In buying mortar-making materials, one cubic foot of mortar can be made from about one cubic foot of sand. To make a portland cement-lime mortar, you'll need one sack of portland cement and a sack of lime for each five cubic feet of sand. To make mortar-cement mortar, you'll need a sack of masonry cement for every three cubic feet of sand.

BRICK PATTERNS

Bricks can be laid in many different patterns that a mason calls *bonds*. A single-tier wall built one brick thick can use either a running bond or a stacked bond. For beginners the running bond is much easier to lay, and is therefore, recommended. It is also stronger. A two-tier eight-inch-thick brick wall needs crosswise bricks called *headers* laid into the wall about every sixth course in height. The purpose

Buttering the ends of bricks so that the mortar will stay on them takes a little practice. It helps to wipe the trowel down across the mortar at an angle from each side of the brick to taper mortar.

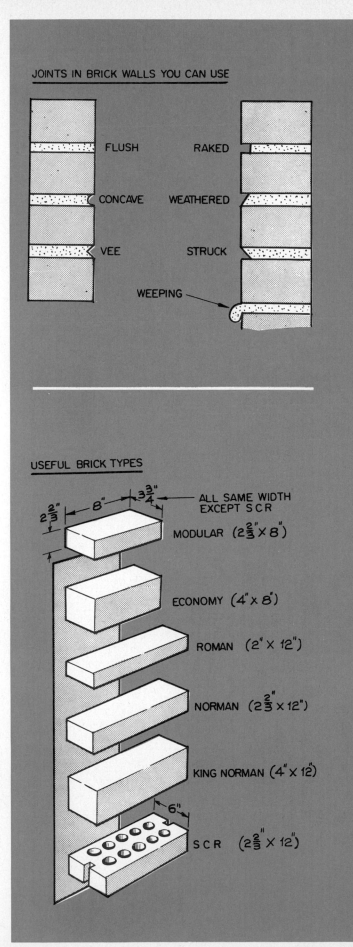

JOINTS IN BRICK WALLS YOU CAN USE

FLUSH   RAKED

CONCAVE   WEATHERED

VEE   STRUCK

WEEPING

USEFUL BRICK TYPES

ALL SAME WIDTH EXCEPT S C R

MODULAR ($2\frac{2}{3}$" × 8")

ECONOMY (4" × 8")

ROMAN (2" × 12")

NORMAN ($2\frac{2}{3}$" × 12")

KING NORMAN (4" × 12")

S C R ($2\frac{2}{3}$" × 12")

Laying bricks from the corners each way goes quickly. Spread mortar for a full bedding. Then butter a brick's end and set it against the corner brick. Tap level and into careful alignment.

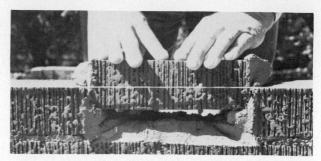

The last brick in a course is called a closure brick. Laying it takes special treatment. Butter the ends of both the opening and the brick. Lower it gently and tap down level to the line.

Varying the way you lay the bricks can make all sorts of patterns in the wall. This one is open to let the breeze come through a brick fence while still giving privacy. Adaptable things, bricks!

of these is to tie the tiers of the wall together all the way up. The simplest way to get them is to place an entire course of header bricks every sixth or seventh course as you go up the wall. This, however, creates horizontal lines in the wall. You may prefer to use one of the Flemish bonds shown in the drawing. These take in some headers into each course, thereby masking their use. Whatever bond you choose, keep its patterns in mind as you lay each brick.

GETTING STARTED

Have your bricks stacked handy to the job and your mortar-mixing operation ready to go before you start bricklaying. Also, make a *story pole*, a long 1x1 or 1x2 stick with the tops of the brick courses carefully marked starting with the first and reaching up to the top course. This saves much measuring.

Bricklaying consists of building up corners, stretching a line between corner bricks and laying in-between bricks to the line. Start the process by laying out the first course of bricks on the footing, dry without mortar. Dry-laying will show you how the bricks will fit what you're building. Make the end joints—the joints between bricks in the same course—to the desired dimension. The thicker joints are recommended because they allow a greater margin for error in brick placement. Thin mortar joints are difficult

to lay accurately. If the bricks don't come out as you wish, you can either adjust the end joints slightly (not much or it will look strange in the finished wall) or you can use part of a brick in the wall, cutting it with a mason's hammer.

If everything fits, mark the centers of the end joints on the footing and take up the dry-laid bricks.

BUILT UP CORNERS

All bricks should be soaked, but allowed to surface-dry before laying them. One way is to sprinkle the pile regularly all day. Another is to keep a soaking pail full of water and bricks, taking them out a few minutes before laying. If not wetted, a brick's bond will be poor because it steals all free water from the mortar.

Start the operation by spreading enough mortar for one brick on top of the footing in one corner. Place the story pole on the footing at that location and lay the brick on the mortar tapping it down with the trowel handle until its top is even with the story pole mark for the first course. Take away the story pole and lay your level on the brick lengthwise. Tap the high end down until the brick is level. Lay your level crosswise on the brick and tap until the brick comes level that way, too. Always tap on the side toward the bubble. Finally check the brick height again using the story pole and again

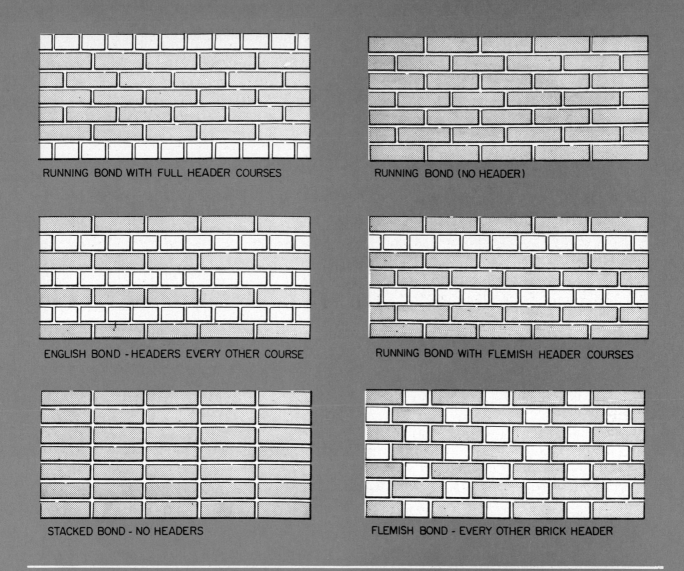

RUNNING BOND WITH FULL HEADER COURSES

RUNNING BOND (NO HEADER)

ENGLISH BOND - HEADERS EVERY OTHER COURSE

RUNNING BOND WITH FLEMISH HEADER COURSES

STACKED BOND - NO HEADERS

FLEMISH BOND - EVERY OTHER BRICK HEADER

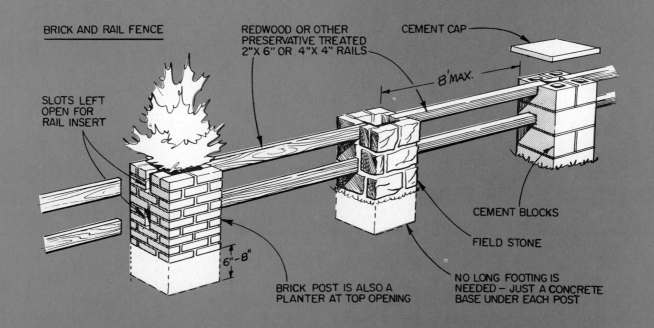

BRICK AND RAIL FENCE

REDWOOD OR OTHER PRESERVATIVE TREATED 2"x 6" OR 4"x 4" RAILS

CEMENT CAP

8' MAX.

SLOTS LEFT OPEN FOR RAIL INSERT

6"-8"

BRICK POST IS ALSO A PLANTER AT TOP OPENING

CEMENT BLOCKS

FIELD STONE

NO LONG FOOTING IS NEEDED – JUST A CONCRETE BASE UNDER EACH POST

Cut bricks by hitting them sharply along a line with a mason's hammer. You can hold the brick in your hand as you strike it, or set it in sand. The same can be done with a trowel, but it's harder.

BRICK AND MORTAR QUANTITY FOR 4" WALLS *				
BRICK SIZES IN INCHES	NUMBER PER 100 SQ.FT. WALL	CU. FT. MORTAR PER 100 SQ. FT. OF WALL		
		JOINT THICKNESS		
* ALLOWS FOR WASTE		1/4"	3/8"	1/2"
2 2/3 X 4 X 8	742	4.0	5.8	7.3
2 X 4 X 12	660		6.7	8.6
2 2/3 X 4 X 12	495	3.7	5.4	6.8
3 1/5 X 4 X 8	618	3.5	5.0	6.4
4 X 4 X 8	495		4.3	5.5
4 X 4 X 12	330	2.7	3.9	4.9
2 2/3 X 6 X 12	495		8.2	10.6
2 1/4 X 3 3/4 X 8	715	4.2	6.0	7.6

BRICK AND CONCRETE STEPS

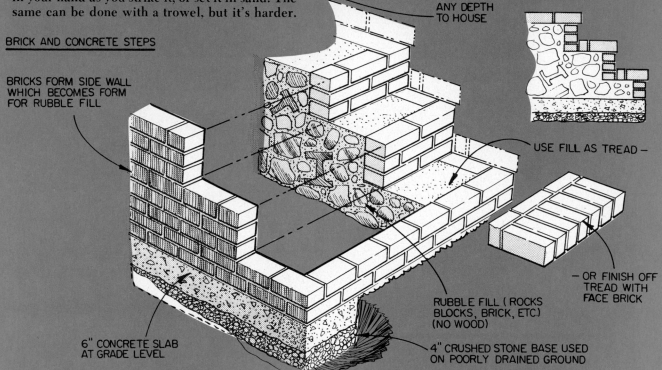

BRICKS FORM SIDE WALL WHICH BECOMES FORM FOR RUBBLE FILL

ANY DEPTH TO HOUSE

USE FILL AS TREAD —

— OR FINISH OFF TREAD WITH FACE BRICK

RUBBLE FILL (ROCKS BLOCKS, BRICK, ETC.) (NO WOOD)

4" CRUSHED STONE BASE USED ON POORLY DRAINED GROUND

6" CONCRETE SLAB AT GRADE LEVEL

check the level. Adjust if necessary. That's a lot of fuss for one brick, simply because it's a corner brick. It controls all the bricks in the course. Get it right and the rest will be right.

Now lay several more bricks stretching out from the corner brick toward the opposite end of the wall. Go only as far as your level will reach. In any case, four bricks is plenty. Place the level on the corner brick and tap subsequent ones down level with it. Use the level placed along the face of the row of bricks to bring all bricks into perfect alignment with each other.

All adjustments must be made while the mortar is soft. Once it dries out, any movement of a brick will destroy its bond. Such a brick should be pulled out, cleaned of mortar, and relaid in fresh mortar.

94 Lay a row of several bricks at the second

corner, aligning them with the level used as a straightedge.

Now build up the corners about four bricks high. If the wall goes around a corner, get started in that direction as well, being sure to maintain your desired bond pattern with the use of overlapping full bricks and half-bricks around the corner.

Instead of headers, metal ties can be used to hold the tiers together. These are embedded in mortar joints and do not show on the wall. There should be at lease one tie for each 4½ square feet of wall area. Stagger the ties in alternate courses.

LAYING IN-BETWEEN

Now lay the in-between bricks to a tight.

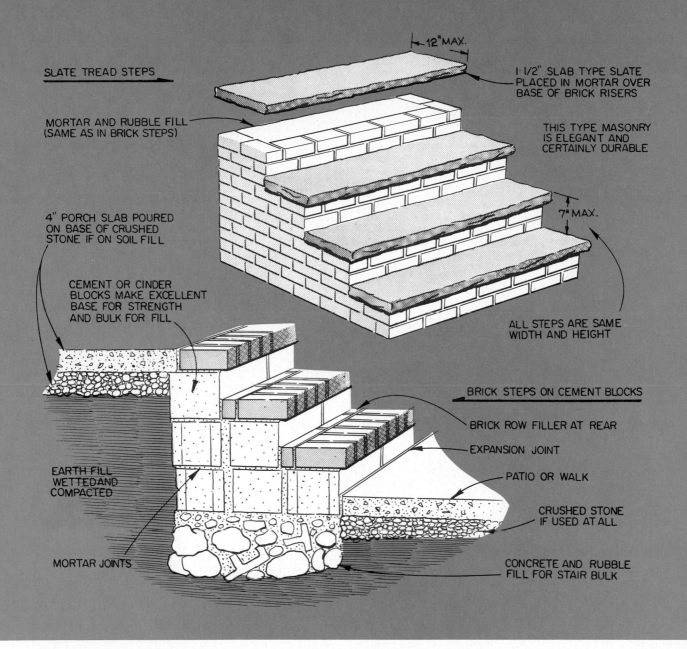

SLATE TREAD STEPS

12" MAX.

1·1/2" SLAB TYPE SLATE
PLACED IN MORTAR OVER
BASE OF BRICK RISERS

MORTAR AND RUBBLE FILL
(SAME AS IN BRICK STEPS)

THIS TYPE MASONRY
IS ELEGANT AND
CERTAINLY DURABLE

7" MAX.

4" PORCH SLAB POURED
ON BASE OF CRUSHED
STONE IF ON SOIL FILL

CEMENT OR CINDER
BLOCKS MAKE EXCELLENT
BASE FOR STRENGTH
AND BULK FOR FILL

ALL STEPS ARE SAME
WIDTH AND HEIGHT

BRICK STEPS ON CEMENT BLOCKS

BRICK ROW FILLER AT REAR

EXPANSION JOINT

EARTH FILL
WETTED AND
COMPACTED

PATIO OR WALK

CRUSHED STONE
IF USED AT ALL

MORTAR JOINTS

CONCRETE AND RUBBLE
FILL FOR STAIR BULK

mason's line. In the time it takes to lay one corner brick, you can lay five in-between bricks. Stretch the line between the corner bricks when the mortar has set enough so they can stand put. Spread mortar on the footing between corners and begin laying the first in-between course. Butter one end of each brick before you lay it. Then, as you set it in the mortar, give it a gentle shove downward and toward the adjoining brick. Tap it into alignment with the string and to the proper distance from the next brick using the mason's hammer. In every case the line should be flush with the top of the brick and about 1/16 inch away from it, not touching it. The bricks should be level, too. Check each one with a level until you get adept at judging what's level.

The last brick in the course is called the closure brick. It gets a joint at each end. Do it

as shown in the photo by buttering both ends of the opening as well as the closure brick.

Cut off excess mortar that oozes from the joints and return it to the mortarboard. As soon as the first course has been completed, move the line up one course. And so on up the wall.

After tooling the joints in a day's work, brush off all loose mortar from the face of the wall with a brush. After a week or more, wet the wall with a hose and scrub it with a 1:9 solution of muriatic acid in water. Use a stiff fiber brush and wooden scraper. Be sure to wear rubber gloves, goggles, and old clothes. The acid will remove all traces of mortar on the wall. Rinse it completely before letting it dry out.

For gray and buff-colored bricks, clean with trisodium phosphate (TSP) and *hot* water. Rinse thoroughly as for acid-cleaning.

The Art of Repointing Bricks

Preparation is the most important ingredient when the task of repointing bricks is called for.

ABOUT THE ONLY MAINTENANCE a brick wall needs is an eventual repointing. This is the renewing of the exterior portion of the mortar joints. After many years the mortar begins to weather. It softens, crumbles, and some of it may even fall out of the joint. If the process continues long enough, all the mortar will become affected and whole bricks may loosen up. Fortunately, in a well-built wall, this takes years.

Repointing consists of chipping out the old mortar an inch or so deep, and replacing it with new mortar. Do the chipping with a cold chisel, a mason's chisel or cape chisel. Since the mason's chisel contacts a wider area, start with it. If you run into tough spots, try using the cold chisel. If the going gets tough, bring the cape chisel to bear. A mortar-raking tool also may be used if the mortar has decayed greatly.

If you have much repointing to do, find a tool rental firm that will rent you a tuckpointer's grinder. It has a Carborundum abra-sive wheel, spinning at 400 rpm to clean out mortar joints to the desired depth—fast. On smaller jobs, you might try using an abrasive wheel chucked into a half-inch electric drill. Chip away at small pieces of the old mortar, aiming the tool back toward the chipped out portion. It's easier that way. Be careful not to damage the bricks themselves. Clean by blowing the joints out with compressed air from a paint sprayer, a vacuum cleaner, or a brush.

POINTING

Tackle the repointing in a ten-foot square area. Dampen the brickwork before you repoint a section.

Always work from the ground or from a sturdy scaffold. (You can rent a scaffold.) It isn't safe to repoint high walls from a ladder.

Mortar should be the same as that recommended for laying masonry (see the chapter on mortar). If you aren't doing the whole wall, but

1 Getting the old mortar out of the joint to a depth of about an inch is hard no matter how you do it. Try various chisels to see which works best, wearing safety glasses as you pound.

2 Dampen the whole prepared wall by sprinkling, then push stiff mortar into each joint. Fill the vertical joints first, next the horizontal ones. Finish by tooling to match rest of wall.

3 Always wear goggles or safety glasses when chipping out mortar. You never know when a chip will shoot toward your eye; the risk is much too great. Glasses are least prone to cloud up.

If the mortar is very soft and aged, a wire wheel chucked in your electric drill can put power to work cleaning out mortar joints. The stiffer the brush you use, the faster and better the job.

Plasterer's hawk makes it easiest to push mortar into the raked-out joints between bricks. Concentric circles on the hawk plate keep the mortar supply from slipping off when the hawk is tilted.

are simply repairing a part of it, try to match the new mortar color to the old by adding a concrete coloring pigment. Mortar, when wet, should look somewhat darker than what you want it when dry. Good repointing mortar is stiffer (has less water) than mortar used for laying brick. It should slide off the trowel when turned sideways, but cling to it when inverted (see photo).

Place the mortar on a hawk. It's a flat board with a handle underneath. Hold the hawk up next to the first joint to be pointed and pack the joint full of mortar with a narrow trowel. This has an offset handle to clear your knuckles from the brickwork. Trowel blades are seven inches long and may be a quarter-inch to one inch wide. Blade width should be no more than the width of the masonry joints.

Fill the vertical joints *first*, then the adjacent horizontal joints. Later, tool the filled joints to match existing joints. Clean up the same way as for new brickwork.

Finish off the repointing job by tooling the joints. Here a short curved length of copper tubing is used to make a slightly concave joint. Time for tooling is when mortar is "thumbprint" hard.

How to Lay Stone

Rubble rock walls need not be straight and level. Curves and sloping tops are easy. Minimum width for load-bearing rubble walls is 16″. Rock walls require more mortar than brick walls.

You can build outdoor projects that will be the making of your house. The inner structure of this massive yard light consists of concrete block. Masonry ties holds the stones in place.

The ultimate in masonry work is being

STONE, often considered the ultimate in masonry beauty, need not be the ultimate in cost. Random stones picked up along the road or found in creek beds have a natural "belongs there" look, unlike man-made materials. However, laying rubble stone masonry is more work and takes more skill than laying brick and block. This is because fieldstone doesn't come in standard sizes and shapes with square corners and flat tops and bottoms. If you buy stones, you can get them squared off enough to permit laying with slim joints like brick and block. Then the procedure is not much different than bricklaying. Wall thickness in that case may be the same as for brickwork. Random stone masonry walls, however, must be at least 16 inches thick if they are load-bearing. They have only half the strength of other masonry walls that are built with squared-off units. Also, rubble masonry joints are wider, ranging from a half-inch to one inch thick.

Just as in brickwork, stone walls should have headers, stones that reach completely from the front to the back of the wall. These bond the facing and backing stones into a strong, structural unit.

Stones, even fieldstone rubble, may be laid either coursed or random. Coursed lay-up calls for more chipping to fit them, but it makes a stronger wall. For maximum strength an equal number of header and stretcher stones should be used in the wall. Avoid, if possible, long vertical joints in the wall.

FITTING STONES

Stone masonry mortar is the same as that used for laying bricks and blocks. *Shape each stone with a mason's hammer before laying it.* The amount of work you put into this step determines how closely the stones fit and how strong the wall is. Also, it governs how formal the wall will look. A wall built with very little

RANGE

able to work with stone—and here is how it can be accomplished.

stone fitting has both thick and thin mortar joints and presents a rustic appearance. That's all right if rustic is what you want and if you build the wall at least 16 inches thick so it has sufficient strength.

Fitting is done with both ends of the hammer, whichever gives the results you're after. Stones will break cleanly to a line chipped around them if you keep chipping along the line with the chisel end of the hammer. Once broken, irregularities in the face can be brought down by swift glancing blows with the head end of the mason's hammer.

To avoid a too-finished appearance in cut stone, a heavy hammer is used to break chunks out of a stone's corners and thus give it a rough look.

LAYING THEM

Because of the roughness of most stonework, you needn't get each cornerstone laid perfectly plumb above the one below. Hold the level in plumb position half an inch or so away from the growing stone wall and make the wall generally plumb at the corners. Peaks of each cornerstone may stick out and almost touch the level, but these are ignored. The main body of the stone wall is what counts in positioning it. This is easy because you don't have to be exact. It's harder, too, because it depends on your having a good eye for alignment before you may have had time to develop one! Still, if you're satisfied with each stone as you position it, you should be well satisfied with the entire wall.

In stonework, a line is loosely strung between corners to help keep the walls straight. Individual units are not laid to the line as with bricks and blocks.

Lay the corners first, no different than other masonry, using a full bedding of mortar.

Larger stones should be laid in the lower portion of the wall; the small ones should be laid in the upper part. This looks right because it's the way of nature. Do it the other way around and the wall will look unnatural and top-heavy.

Spread mortar and lay each fitted stone in place among others in the growing wall. Chips of stone may be placed in the mortar to hold the stone from sinking out of position while the mortar sets. In any case, the stone should end up with a full bedding of mortar around it.

Break up stones that are too long for their height and width. Strong stones like granite can be up to five times as long as they are wide. Limit this to three times for soft stones such as limestone and sandstone.

If you're building a rough rubble wall and find that mortar joints are hard to keep filled with mortar to a uniform appearance, you can rake them all out an inch deep. Later, the joints can be filled with stiff mortar to the desired depth. This depth should leave the joints no more than an inch wide. Slim joints make your wall look more stone-like, less mortar-like.

HOW MUCH MORTAR?

Figuring mortar quantities is more of a gamble with stones than it is with conventional masonry. A good rule-of-thumb for squared-off-stone is to allow seven per cent of wall volume —volume not area—for mortar using half-inch joints. Allow twice that if the stones are small or only roughly squared. For rubble masonry, allow up to 35 per cent of wall volume in mortar.

Since an acid wash cannot be used to clean stonework because it stains the stones, be extra careful to keep mortar off the faces of the stones. Remove mortar stains when you tool the joints, by using a brush. If stains are stubborn, try using a wire brush.

BROKEN RANGE (ASHLAR)

RANDOM RUBBLE

Some experts recommend cleaning the stones with a laundry detergent and warm water. First wet down the wall with a garden hose, then scrub the detergent in with a brush. Rinse and wash again. Finally rinse thoroughly. Sometimes a jigger of household ammonia is added to the wash water. At any rate, don't clean the stones with an acid solution.

Stones that are flat and layered—slate for example—should always be laid in their natural

Rocks are there for the collecting, along roads, and in fields. Ask the land owner for permission first, then toss them into a pickup. Hand-pick as you go, hosing them clean before unloading.

Start laying cleaned, shaped rocks into their mortar bed. For best appearance use the larger rocks at the base of the wall. Use rubber-dipped or leather-faced gloves to protect your hands.

position—flat. If you lay them vertically, the wall will not look right and its strength will be adversely affected.

The amount of wall you can build with a ton of stones varies depending on the material, its degree of finish, and the wall thickness. However, most rubble stones will build from 25 square feet to 50 square feet per ton. Most cut stones will make around 50 square feet per ton. Selected finely cut stones can cover more than 100 square feet per ton. Prices run from $35 to $400 a ton, those with better coverages usually costing more. Local stones, of course, cost less than stone hauled from some far-off quarry. Slate, flagstone, and other stones used for paving and flooring, will of course cover greater areas.

Outside of building with rammed earth, stone is the most natural, unprocessed-by-man material you can use. You can find it in many

Rock-work should be laid in a full bedding of mortar, both on the footing and above. Mortar should be slightly stiffer than that for bricklaying. Limit height to 2' per day to avoid collapse!

Just as in laying bricks and blocks, tap each rock in place before setting the next one. If the wall is more than one rock thick, some larger rocks should laid crosswise wall width for strength.

places, free for the collecting. Many farmers and ranchers would be happy to let you carry off stones from their fields, *if you ask them.* Likewise, stone quarries often consider the rubble left over after selling the "good" stones to be worthless. For a small cost they might let you take what you want and haul it away. Stratified stones like limestone can often be brought down to lifting size by hammering steel wedges into the natural cleavage lines. You can also buy the stones you need, rubble or dressed to any degree you like.

RUBBLE

Rubble is raw, unprocessed stone as you find it. It makes fine-looking walls. When laid with straight horizontal joints as in brick and block, it's called *coursed rubble.* Laid without regard to pattern, it's called *random rubble.*

Keep this in mind, too, as you lay stones: stones should work together for size, shape, and color. That is, they should look as though they belong together. There may be a variation in color, but keep it within the natural variation found in nature. Moreover, two stones of the same size and shape should not be laid side-by-side. Separate them with either smaller or larger stones. For strength, stones should extend at least one-third of the way into the wall. In a 16-inch-thick wall no stone should be less than five inches in width. Always lay stratified stones with their cut edges on the face of the wall. The striations if exposed, to the weather, will bring water into the wall's interior, sometimes with disastrous results.

All stones should be dampened before being laid in the wall. It's a good idea to keep them damp by sprinkling for a week after they're laid.

Spread mortar for the next rock on the end of the one just laid, after fitting the rock to its spot by chipping(if necessary) with a mason's hammer. Keep the joints no more than 1" wide.

Set the next rock in place, pressing it into the mortar. In rock-work it is permissible to prop rocks into the desired position with chips of stone. The mortar must surround each rock.

Corners of rubble rock-work are merely "aimed" at being plumb. Since the irregular shapes of of all rocks cannot be plumbed like regular bricks and blocks, they need only look plumb.

Since keeping full mortar joints in rough rubble masonry isn't always practical, the joints may require pointing after laying. First, rake the joints, then fill the cavity to the desired depth.

Make a Patio

Making a patio a livable place is like adding an extra family room to your house—at less than half the cost.

The best patios give a sense of complete enclosure, hence the brick wall around this one. The edge of the slab was thickened and steel reinforced to 12" to form what is called a grade beam to support wall.

AN EXCELLENT PATIO for outdoor enjoyment can be made of concrete, bricks, precast concrete tiles, and many other materials. Combinations of these can be used, too. The most popular patio material is probably concrete. It can be poured to any desired shape, is practically maintenance-free, and can be colored and textured to suit.

A patio should be below the interior house floor level to keep water from coming in under the door. A one-inch difference is enough. Of course, you can have more than an inch difference if you wish. In fact, a patio can be made on various levels to take advantage of sloping ground. The patio should also slope away from the house about ¼ inch per foot to carry off rain and melting snow.

A concrete patio is usually built four inches thick with control joints dividing it into ten-foot maximum size slabs to prevent cracking. The subgrade is prepared the same as for a sidewalk or driveway. See those chapters for details. Forming is similar, too. A sand or stone surface, if used, should extend a foot beyond the patio edges to prevent undercutting in a heavy rain. Provide some method of draining to keep water from collecting along the edges.

To save on concrete you can build a patio only 1½ inches thick, *if* it is jointed into two-foot maximum size slabs to prevent cracking.

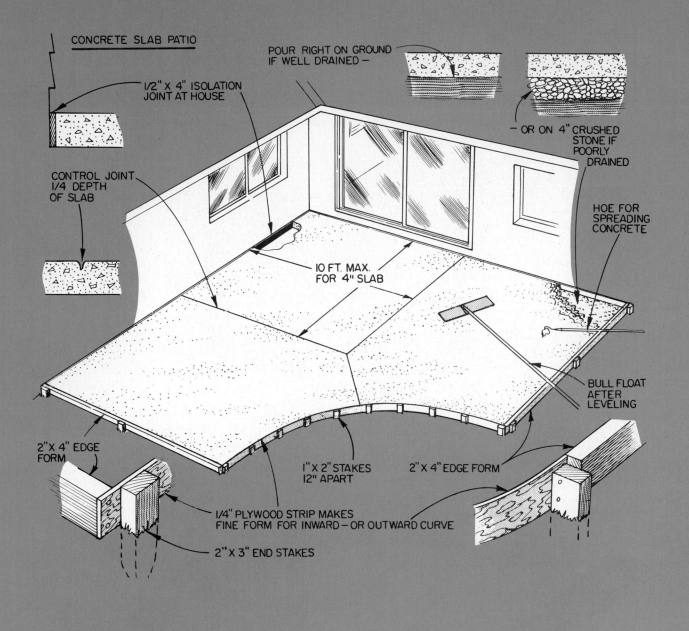

CONCRETE SLAB PATIO

1/2" X 4" ISOLATION JOINT AT HOUSE

POUR RIGHT ON GROUND IF WELL DRAINED —

— OR ON 4" CRUSHED STONE IF POORLY DRAINED

CONTROL JOINT 1/4 DEPTH OF SLAB

HOE FOR SPREADING CONCRETE

10 FT. MAX. FOR 4" SLAB

BULL FLOAT AFTER LEVELING

2" X 4" EDGE FORM

1" X 2" STAKES 12" APART

2" X 4" EDGE FORM

1/4" PLYWOOD STRIP MAKES FINE FORM FOR INWARD — OR OUTWARD CURVE

2" X 3" END STAKES

One way to do this is with a jointing tool, as described in the chapter on tools. Another way to form the patio into small squares or rectangles is with non-rotting 1x2-inch redwood, cypress, or pressure-treated wood. The frames can be partially prenailed and assembled in ladder-like strips. If you stagger the adjoining wood form strips, all can be end-nailed. Lay them on the subgrade and stake them temporarily. Level, if necessary, by sliding blocks under the low spots. Since the thickness is only 1½ inches, a load of ready-mix builds a lot of patio.

To prevent the strips from working up or down and spoiling the future appearance of your job, drive 8-penny galvanized nails through the strips at frequent intervals. These will become embedded in the concrete tieing it and the strips together.

PREPARATION OF SUBBASE

With a thin-section patio, subbase preparation is important. Tamp it well, or else cast the patio on unexcavated earth. Getting that to proper grade is tough, so a sand fill is recommended. Dampen and compact it. Use a home-built hand tamper made by nailing an eight-inch-square of plywood to a length of 4x4. If

Rented tractor with rear scraper blade made fast work of leveling a patio slab for do-it-yourselfer Chuck Kearns. It works only if you can get the rig in and out; otherwise use the shovel.

Ready-mix for Ernie Cowan's patio was chuted in over a fence. To get the average mixer in, you'll need a path 8-9' wide with 11-13' overhead clearance. Ask dealer for clearance data.

you have a large area to do, you may want to rent a power compactor.

An isolation joint should be used at the house wall, at the sidewalk, the driveway and at other adjoining walls and slabs. This is described in the chapter on sidewalk-building.

NO-MORTAR BRICK OR TILE

A serviceable patio can be made just out of bricks or precast concrete tiles set on a sand bed. First, construct a non-rotting wood edging or a trench with a row of bricks set on end for an edging. An edge is needed to hold the bricks or tiles together. Dig the sod out down to a level that will place the tops of the units at the desired grade. Set the edging with its top at or above finished patio grade. Then spread a two-inch layer of sand and smooth it to slope away from the house ¼ inch per foot.

Now you can begin laying the units. Choose any pattern you like, combining different-sized precast concrete tiles, paving blocks, or bricks. Bricks you use should be hard-burned bricks that will be able to take the severe weathering they'll be subjected to on the ground. Besides bricks, a fine no-mortar flagstone patio can be laid the same way.

Below, leave-in pressure-treated wood (or untreated redwood, cedar, cyprus) work has been positioned, ready for pouring. No other jointing is needed, since the forms make their own.

Above, curing fresh concrete exposed-aggregate patio squares by covering the area with newspapers, then keeping wet for six days makes them strong, prevents later loosening of stones.

Below. Just before pouring concrete on it, the subbase should be throughly dampened with water. This keeps the subbase from absorbing water from mix. Puddles mean too much water.

Above, sand subbase is leveled off below the tops of the forms with a 1 x 2 "T" strikeoff screed. Upper 1 x 2 rides the top of the form while the other one extends down to screed off excess sand.

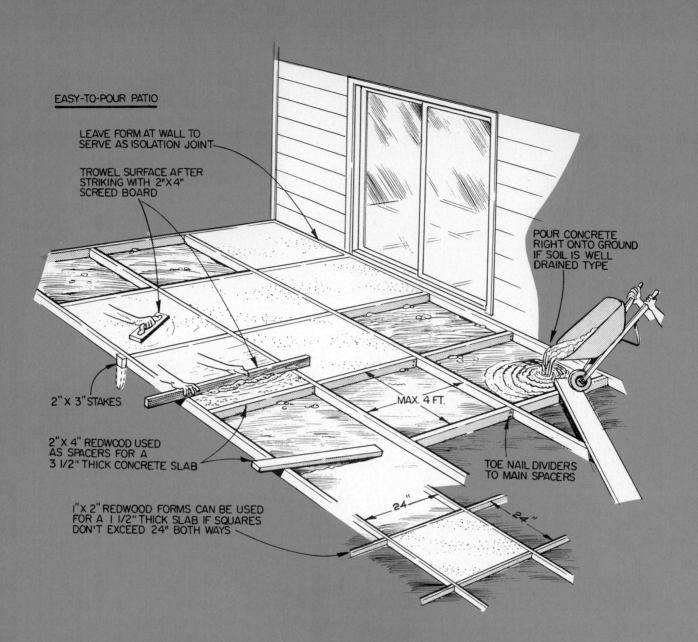

EASY-TO-POUR PATIO

LEAVE FORM AT WALL TO
SERVE AS ISOLATION JOINT

TROWEL SURFACE AFTER
STRIKING WITH 2"X 4"
SCREED BOARD

POUR CONCRETE
RIGHT ONTO GROUND
IF SOIL IS WELL
DRAINED TYPE

2"X 3"STAKES

2"X 4" REDWOOD USED
AS SPACERS FOR A
3 1/2" THICK CONCRETE SLAB

MAX. 4 FT.

TOE NAIL DIVIDERS
TO MAIN SPACERS

I"X 2" REDWOOD FORMS CAN BE USED
FOR A I 1/2" THICK SLAB IF SQUARES
DON'T EXCEED 24" BOTH WAYS

24"

24"

This patio area was made with
cinder patio bricks measuring
4"x8"x16". The area was first
leveled and tamped, then 2" of
sand was spread. Just before laying
a group of four blocks the sand was
sprinkled with portland cement
and the blocks set level. After the
patio was completed the area was
hosed down. Water seeped through
the cracks and the cement and
sand set to give each block a
firmer base. Many difference sizes
and colored blocks are available.

Whatever material you lay, throw damp sand over the finished surface and sweep it into the cracks between until the sand is flush with the surface. Often the wider joints between flagstones are filled with topsoil or sod to get "living" joints between the flags.

If a wood edging is used, you can make a self-leveling strikeoff board that, drawn over the tops of the forms, will smooth off the sand bedding to the right level and slope (see drawing).

CONCRETE AS A BASE

Bricks, tiles, and flagstones may be set on a solid concrete base for a patio surface that needs almost no maintenance. First pour the concrete base almost as though it were a patio itself. Make it three inches thick, float it, and scratch the surface with the edge of a piece of hardware cloth or chicken wire to provide "tooth" for what's to follow.

Some method must be devised to prevent shrinkage cracks that are bound to form in the concrete base from affecting the tiles, bricks, or flagstones. The best way to build the base is with 6x6-inch No. 10 steel mesh laid in the halfway section of the three-inch base slab. This won't prevent cracking, but it will hold the cracks tightly together and, hopefully, keep them from cracking the upper masonry units. Another method is to form control joints in the base slab no more than ten feet apart and match them with the joints above. This is a bit harder to work out. You need to know exactly where the top joints will fall. This is a sure way to avoid unsightly cracks.

The following day spread a one-inch layer of mortar (see the chapter on making mortar) over a portion of the base slab. Begin laying paving material in it right away. Pound them level and align with a mason's hammer, being careful not to chip the top surfaces and edges. A long level or straightedge will help.

USE OF ADHESIVES

You can wait as long as you like between building the base slab and preparing the base concrete for a good bond, if you apply a good vinyl concrete adhesive or a similar bonding agent (see the chapter on repairs). You can also use a bonding slurry made of portland cement and water. Brush it into the pores of the base slab and place the mortar topping while it is still damp.

Stay off the patio for one day while the

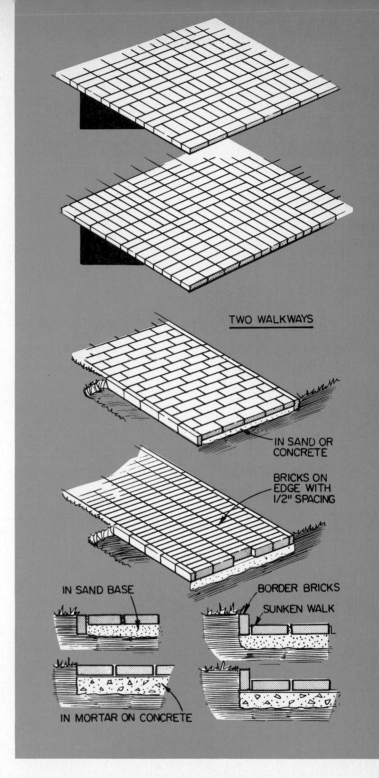

TWO WALKWAYS

IN SAND OR CONCRETE

BRICKS ON EDGE WITH 1/2" SPACING

IN SAND BASE

BORDER BRICKS

SUNKEN WALK

IN MORTAR ON CONCRETE

mortar sets up. If you must walk on it to work, lay sheets of ¾-inch plywood or planks to work on.

The next day, fill in the joints. For an antique effect, you can let the mortar hang over the edge joints. For a formal effect make neat mortar joints, as in a brick wall. A long, narrow trowel made for tuck-pointing may be helpful here. Within several hours remove mortar stains with a brush. Later, get rid of all stains completely following the directions in the chapter for laying brick, concrete block, or stones, whichever material you are paving with.

Fireplaces and Fire Pits

If you can follow simple instructions, you too can construct a fireplace or fire pit to add immeasureably to the comfort and value of your home.

A FIREPLACE and fire pit may be made of bricks, concrete blocks, or stones. As long as the flue size and the fireplace opening dimensions given in the table are adhered to, it will draw well. The masonry in contact with the fire is made of firebrick laid in fireclay mortar.

You can incorporate a fire pit into a new patio. It provides a place to build a fire for warmth or cooking and serves as an outdoor entertainment center. The fire pit may be raised above the ground with a ring that also serves as a seating area, or it may be sunk into the ground and the seating handled with informal chairs or benches. The simplest fire pit is merely an enclosure for building a fire with a circular or rectangular wall plus some gravel in the center. More elaborate fire pits inclúde a provision for bringing fresh air to the area beneath the fire. The drawings show a variety of types. Draft tubes could be built into any of them to improve the fire's performance by setting six-inch tiles below ground leading from outside the fire pit to the fire area.

ADD-ON FIREPLACE

Building a fireplace in an existing house is a little bigger undertaking. It must fit the house, extend through the roof, and meet local codes. Check them. Your building officials may have a specification sheet on add-on fireplaces that will help you. Their requirements, rather than those shown in our plans, should be followed.

If your house has a concrete slab floor, you can build the fireplace right on it. If your house has a wood floor, make the fireplace with a cantilevered raised hearth to avoid cutting through the floor. The cantilevered hearth also may be used with a slab floor fireplace.

PREFAB FIREPLACE

Decide whether you want to use a prefabricated steel fireplace unit or build your own firebrick firebox. If you want the fireplace for heat, get the prefab unit. Otherwise, forego it to save money. If you use the unit, follow the step-by-step drawings that come with it. Either way, you'll need a building permit.

Plan for your finished chimney height to be at least two feet above the highest part of the roof that's within ten feet. Minimum chimney height should be 15 feet above the hearth. All parts of the fireplace must have a minimum clearance to framing of two inches inside the house and one inch from the outside wall. Wall and floor facing materials may actually touch

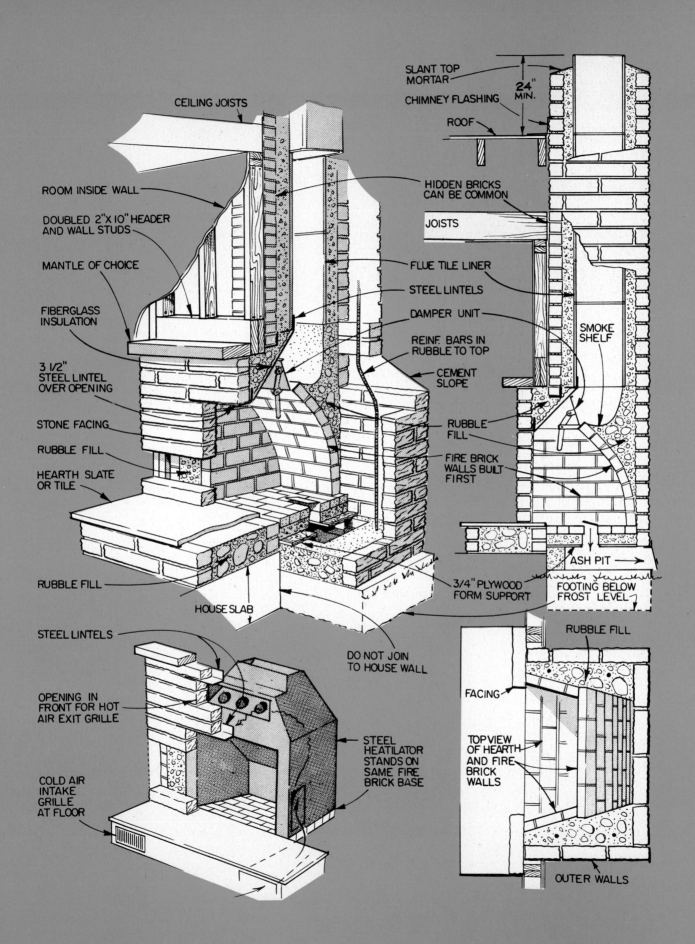

CEILING JOISTS

ROOM INSIDE WALL

DOUBLED 2"x10" HEADER AND WALL STUDS

MANTLE OF CHOICE

FIBERGLASS INSULATION

3 1/2" STEEL LINTEL OVER OPENING

STONE FACING

RUBBLE FILL

HEARTH SLATE OR TILE

RUBBLE FILL

HOUSE SLAB

SLANT TOP MORTAR

CHIMNEY FLASHING

ROOF

24" MIN.

HIDDEN BRICKS CAN BE COMMON

JOISTS

FLUE TILE LINER

STEEL LINTELS

DAMPER UNIT

REINF. BARS IN RUBBLE TO TOP

CEMENT SLOPE

RUBBLE FILL

FIRE BRICK WALLS BUILT FIRST

SMOKE SHELF

ASH PIT

3/4" PLYWOOD FORM SUPPORT

FOOTING BELOW FROST LEVEL

DO NOT JOIN TO HOUSE WALL

STEEL LINTELS

OPENING IN FRONT FOR HOT AIR EXIT GRILLE

COLD AIR INTAKE GRILLE AT FLOOR

STEEL HEATILATOR STANDS ON SAME FIRE BRICK BASE

RUBBLE FILL

FACING

TOP VIEW OF HEARTH AND FIRE BRICK WALLS

OUTER WALLS

Prepare a sound concrete footing of an add-on fireplace just outside the wall where it's to go. Reinforcing steel should be embeded in it as illustrated in photos. Dry-lay first brick course.

Cut out the house siding, studding and interior wall for the fireplace. The opening should be about 64″ wide and 88″ high and framed around it as described in the text. Dimensions can vary.

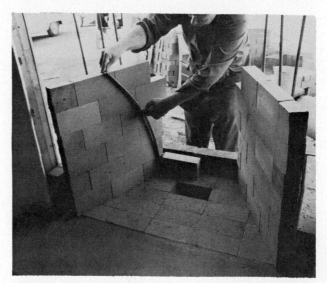

Build up the outside corners to the dimension of your footing. Lay five or six courses and then move to the inside firebricks. Fill in between with mortar, rubble stone or brick chips.

Curving rear wall of the firebox can be marked by bending a thin stick to a smooth curve then drawing its line on each firebrick. Remove bricks and cut to line. Use a hammer and cold chisel.

the fireplace but their framing must have proper clearance.

FOOTING

Build the footing first. Make it below frost depth, one foot thick and six inches wider than the fireplace exterior (54 inches is about the minimum width for a fireplace, but you can build yours as wide as you wish). If reinforcing

bars are required to meet earthquake codes, install these before pouring the footing concrete. The footing should meet the house foundation but not be tied into it. Use a strip of expansion joint material between the two.

WALL OPENING

Cutting the opening through your house wall is a critical step. Have all the fireplace ma-

Lay the firebricks in fireclay using ¼" joints. When the firebox is finished, you can begin laying the interior masonry; here attractive concrete slump blocks. You can use brick or stone.

Purchased metal damper unit (Superior) rests atop the firebox. Rubble masonry (laid anyway) is placed around it. A blanket of insulation will protect masonry from cracking due to expansion.

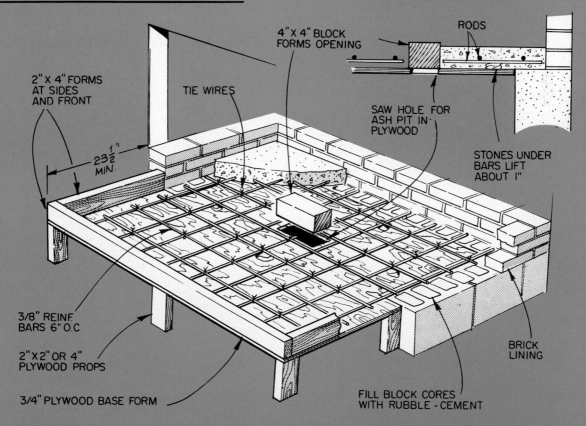

FORMING FOR CANTILEVERED CONCRETE HEARTH

2" X 4" FORMS AT SIDES AND FRONT

TIE WIRES

4" X 4" BLOCK FORMS OPENING

RODS

23½" MIN.

SAW HOLE FOR ASH PIT IN PLYWOOD

STONES UNDER BARS LIFT ABOUT 1"

3/8" REINF BARS 6" O.C

2" X 2" OR 4" PLYWOOD PROPS

3/4" PLYWOOD BASE FORM

FILL BLOCK CORES WITH RUBBLE - CEMENT

BRICK LINING

terials on hand before you start it. Ideally your fireplace should be located in a non-load-bearing outside wall. If not, you'll have to brace up the roof above it by installing temporary 4x4s. Most walls under roof eaves are load-bearing; most walls under gables are not, except perhaps at their centers. Under a truss roof even the center of the gable end wall is non-load-bearing.

Make the wall opening ten inches wider than the outside of the fireplace and about 70 inches above the finished fireplace floor. With a raised hearth, typical cutout dimensions might be 64 inches wide by 88 inches high. Cut through the siding, sheathing, framing, and inside wall material, whether plasterboard or paneling. Remove still plate at the bottom of the opening.

Now install doubled 2x4 framing in both sides of the opening, making use of any framing that is already there. The framing should end up flush with the opening. The exposed 2x4 at each side of the opening should be 9½ inches shorter than the opening. Doubled 2x10-inch headers span the top of the opening and are supported by the shorter 2x4s. If fireplace width is increased much over 60 inches, use 2x12 headers and add two inches to the opening height. Space the headers with 3/8-inch plywood strips and nail them together with 16-penny nails every 16 inches along the top and bottom. Use 16-penny nails for all framing work.

The cut-off wall—called the cripple—should meet the headers across the top. Toenail each cripple to the headers from both inside and outside using two 16-penny nails to each cripple.

ROOF OPENING

Decide whether you want the chimney to be smaller than the fireplace or to stay the same width all the way up. A smaller chimney saves on materials and labor; a full size makes for a massive, luxurious appearance. In any case, the roof framing should be cut back one inch on all sides from the chimney. Use a plumb bob to bring the fireplace corners up vertically from their starting points on the footing. Saw carefully. The cuts you make will show from below. If the roof overhang is great enough, the fascia board can be left intact. If it comes closer than an inch to the chimney, you'll have to remove it. New framing may have to be nailed to the old around the roof opening to brace it.

Setting a flue tile in mortar on the one below is done gently to retain as much mortar as possible. Once through roof cutout, aluminum flashing is laid in joints. Use counter-flashing when roofing.

Capping out the chimney is the end of a job well done. Mortar cap is poured and smoothed with a trowel. Other chimney caps can also be made with plywood forms and then poured concrete.

STARTING THE BASE

Now you're ready to lay the masonry, beginning with the first course laid on the footing. See the chapter dealing with brick, concrete block, or stone for details of working with these materials. The fireplace is made of an outside masonry shell, plus an inside firebox, plus interior rubble fill around a clay floor tile liner. The drawings will help you visualize its construction.

Ash dump and cleanout doors, as well as other metal and masonry parts needed for building a fireplace are available at many building supply dealers specializing in masonry supplies. The cleanout door frame is laid in mortar just as though it were a masonry unit.

The fireplace masonry shell below the firebox floor is left empty for accumulation of ashes between cleanouts. Above the firebox floor, the shell should be filled with masonry rubble consisting of washed rocks, broken bricks, leftover pieces of masonry, cutoffs, etc., and mortar. Looks do not matter.

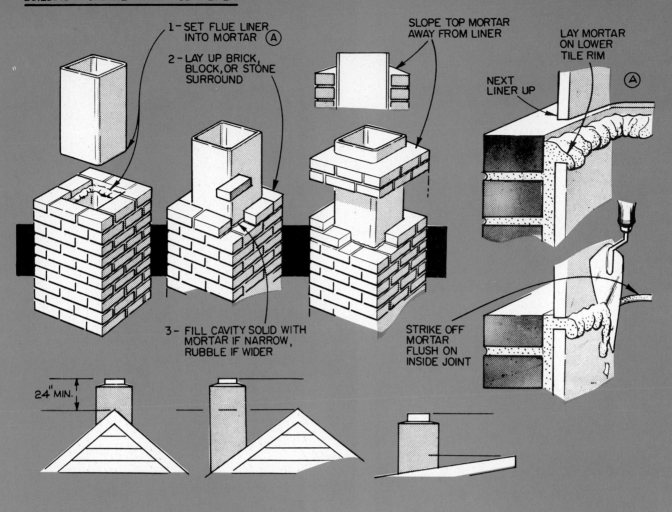

1- SET FLUE LINER INTO MORTAR (A)

2- LAY UP BRICK, BLOCK, OR STONE SURROUND

3- FILL CAVITY SOLID WITH MORTAR IF NARROW, RUBBLE IF WIDER

SLOPE TOP MORTAR AWAY FROM LINER

STRIKE OFF MORTAR FLUSH ON INSIDE JOINT

LAY MORTAR ON LOWER TILE RIM

NEXT LINER UP

(A)

24" MIN.

MAKING THE FIREBOX

If you build a raised, cantilevered hearth, stop laying masonry when you reach about 14 inches above house floor level. Lay an inner tier of concrete blocks or rubble bricks inside the embryonic fireplace shell from the footing up to meet the other masonry. Neatness won't count here. Cut a sheet of ¾-inch plywood to fill the opening in the fireplace shell and extending into the room 24 inches or so from the room interior. This will give you the required 18-inch hearth. The hearth should extend 12 inches beyond the fireplace opening on each side. Lay fireplace shell masonry at least four inches higher, except where the plywood projects into the room. Finally nail 2x4s at the sides and the front of the plywood to form a cast-in-place concrete hearth. Pour it, strike it off, and finish it as smooth as you can. Later, if you wish, the hearth can be covered with tiles. A mat of 3/8-inch reinforcing bars spaced six inches apart can be assembled, and placed in the center of the hearth slab as it is poured. Be sure to make a wood form for the ash dump and position it on the form before pouring. Later you can cut an opening through the plywood to gain access to the area below. Install an ash dump unit in the opening.

MAKING THE HEARTH

Complete the fireplace shell up to the top of the wall opening. At the sides the shell masonry should come into the wall and reach within a half-inch of the finished room wall (but be an inch away from the wall framing).

Now build the firebox. Use standard firebrick laid in a ¼-inch layer of fireclay mortar. This is made of one part portland cement, one part fireclay and three parts damp, loose mortar sand. Mix the fireclay and water first until free of lumps. Lay the firebox floor 20 inches deep using any firebrick pattern. Include the ash dump as one unit. The sides and rear walls of the firebox are made with firebricks laid on their sides in fireclay mortar. Make eight

113

Brick firepit built in an opening left for it in the patio has small holes in its mortar joints to provide draft. This unit was sized to fit the circular grille. Firepits provide great party atmosphere.

courses, cutting the rear sidewall bricks off to allow the rear wall to curve forward five inches from bottom to top. Start the sidewalls so that they are narrower at the back by about five inches. Firebricks used to build the rear wall needn't be cut. They may protrude into the cavity, which is later filled with rubble masonry. Make the firebox so that it projects through the wall opening and comes flush with the sides of the outer shell, that is, a half-inch back from the finished room wall. Fill with rubble and mortar all cavities between the firebox and the shell.

Near the firebox top, begin forming a curved downdraft shelf (see drawing) to divert the air coming down the chimney away from the firebox. Buy and place a damper over the top of the firebox supported by the sides and rear wall of the firebox. Lay fiber glass wool over the damper unit to keep the masonry from coming into rigid contact with the damper. This permits expansion of the heated metal without cracking your masonry. Local codes may require an inch steel lintel (angle iron used to support masonry above it) across the front of the damper unit. Lay rubble masonry over the front of the damper making the courses wider as they approach the top of the damper to follow the taper of the damper front. Lay more masonry on the fireplace shell to bring it up level with the top of the damper unit. The lintel fits just across the top front of the damper unit so bring the interior rubble fill up to that level behind the damper.

Now you're ready to place the first flue

tile. It should be sized to match the cross-sectional area of your fire-place opening. The flue tiles may be any shape. You'll need enough of them to reach from the damper to the chimney top. Spread mortar and set each one carefully. Then reach down inside with your trowel and scrape off the excess mortar. Build up the fireplace shell around the first flue tile before setting the next tile. Bring the rubble fill up along with the shell and liner.

A few inches above the roof, install strips of aluminum flashing halfway into the mortar joints. Bend them down and under the shingles, or else use additional aluminum for counterflashing. These should fit beneath the roofing and extend up about four inches. Then the chimney flashing is bent down over the counterflashing for a rain-tight seal. Seal any openings with plastic roof cement.

The chimney should be topped off when it reaches the required height. Go higher than required if you must for appearance, but never stop too low if you want your fireplace to draw well. You can make a smoothed mortar cap—easiest—tapering it up to the flue liner which should protrude a couple inches above it. Or you can form and cast a sloped concrete top.

FINISHING INSIDE

Inside the house remove the exposed portion of the hearth form and discard it. Tile the hearth if you prefer. Lay up the fireplace masonry across the front. It should lap over the firebox by about an inch on each side so that the ends of firebricks aren't exposed. If you wish, the interior masonry can consist of bricks or cut stone for appearance's sake. At the top of the firebox or somewhat lower, depending on how high you want the fireplace opening to be, place a lintel across the interior course on each side and then lay the next course clear across the front of the fireplace. These courses, may go up as far as you like, even up to the ceiling. If you extend them higher than the cutout wall opening, be sure to install crimped metal wall ties, every two feet, both horizontally and vertically.

Clean up and install trim moldings wherever needed to hide joints between house and fireplace. Install a fireplace screen, mantel and whatever else you want for the face of your fireplace.

Experts advise waiting a week before building your first fire to give the mortar and masonry time to set fully. Make the first fire a small one to help dry the masonry slowly.

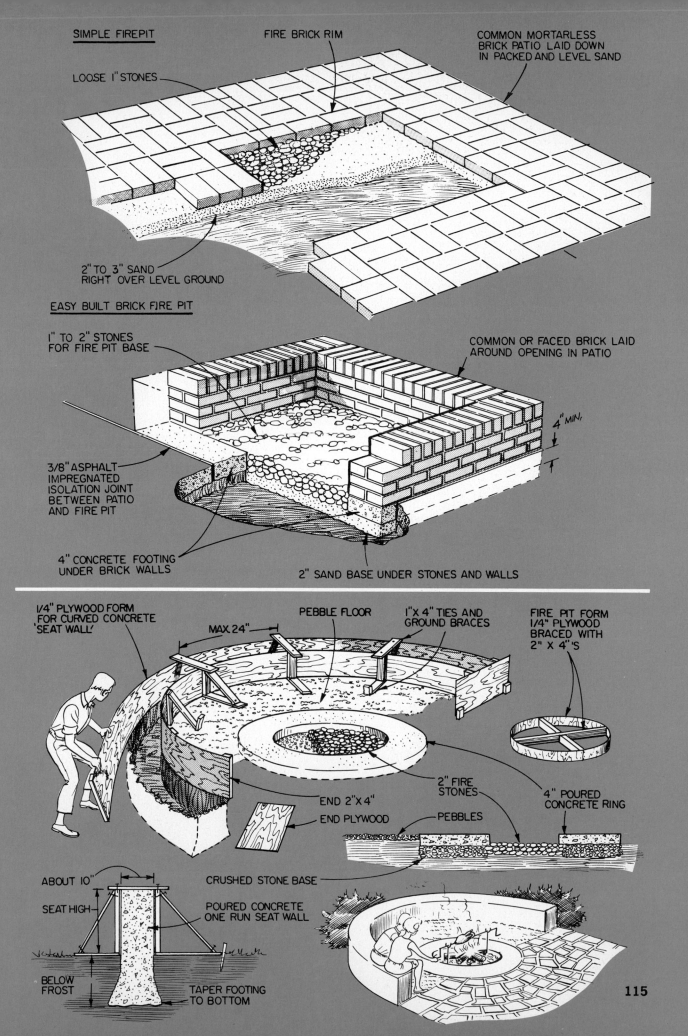

SIMPLE FIREPIT

FIRE BRICK RIM

COMMON MORTARLESS BRICK PATIO LAID DOWN IN PACKED AND LEVEL SAND

LOOSE 1" STONES

2" TO 3" SAND RIGHT OVER LEVEL GROUND

EASY BUILT BRICK FIRE PIT

1" TO 2" STONES FOR FIRE PIT BASE

COMMON OR FACED BRICK LAID AROUND OPENING IN PATIO

4" MIN,

3/8" ASPHALT IMPREGNATED ISOLATION JOINT BETWEEN PATIO AND FIRE PIT

4" CONCRETE FOOTING UNDER BRICK WALLS

2" SAND BASE UNDER STONES AND WALLS

1/4" PLYWOOD FORM FOR CURVED CONCRETE 'SEAT WALL'

MAX 24"

PEBBLE FLOOR

1"X 4" TIES AND GROUND BRACES

FIRE PIT FORM 1/4" PLYWOOD BRACED WITH 2" X 4"'S

END 2"X 4"

2" FIRE STONES

END PLYWOOD

PEBBLES

4" POURED CONCRETE RING

ABOUT 10"

CRUSHED STONE BASE

SEAT HIGH

POURED CONCRETE ONE RUN SEAT WALL

BELOW FROST

TAPER FOOTING TO BOTTOM

115

Cement Plastering

Troweling cement-mortar onto a masonry wall not only changes its texture and gets rid of joint show-through, it helps waterproof the wall. Applications should be in ¼" to 3/8" layers, two layers thick for waterproofing. To help the first layer bond well, the wall should be dampened and coated by brushing on a water-cement paste.

Concrete need not be drab or strictly utilitarian. Here are eight different ways to glamorize concrete.

CEMENT-PLASTERING can help you to hide the mortar joints in a masonry wall you've built. It can also be used to help waterproof the outside of a concrete block or cinder block basement wall. (Bentonite panels described in a previous chapter do a better job). You could also apply a sand-cement coating—called stucco—to the outside of a masonry wall above ground, simply for a smooth appearance. Either way, it involves troweling on a layer of mortar onto the vertical wall surface.

BAGGING-IN

One of the easiest treatments to hide the texture and most joints in block-work is to *bag in* the wall. This fills the joints and closes the block pores with a sand-cement mortar. Tools required are just a soft brush and a burlap bag or a rubber float of the kind used for some concrete finishes.

Make the bagging-in mixture with 1 part portland cement and 1½ to 2 parts mortar sand. Mix dry, then add enough water to make the mix fluid but not runny. Dampen but don't soak the blocks. Use a fine spray from a garden hose nozzle. Apply the mix with the brush, working it into the block pores. Do about five square feet at a time. After applying, go back over the surface with the dampened burlap bag

or rubber float and rub off the excess mix up to the high spots on the wall. This should leave a thin, masking coat over the blocks. The rubber float generally gives a finer texture than the burlap.

Dampen the wall again as soon as it's able to take it without washing off the mix. Keep it damp for 48 hours.

STUCCO

A coat of stucco has the same effect as bagging-in but it is thick enough to completely hide the texture beneath. First dampen the wall. Then apply a 1 to 2½ portland cement mortar sand mix from a plasterer's hawk, using a cement-finishing trowel. You can also use regular masonry mortar. Hold the hawk piled with mortar next to the wall and angled away from it. Trowel the mix upward onto the wall. It should be spread about ¼- to 3/8-inch thick. Once applied, it can be smoothed by further troweling or texturing. Texture with the trowel, a rubber float, or almost anything that produces an effect you like. Interior surfaces should be smoother than exterior ones. Tin cans, cookie cutters, hand tools, brushes, all make attractive designs in the soft stucco. Cure the same as for the bagged-in coating.

Plain concrete block planter is decorated with two coats of troweled-on cement-plaster. The first coat was troweled smooth. The second coat was spotted, for texturing, then troweled lightly.

Other methods of texturing cement-plaster are numerous. This one is done by pressing the corner of a trowel into the fresh mortar coat. The marks could be subdued by light troweling.

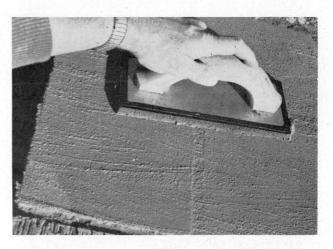

Bagged-in finish is started by brushing the bagging-in mixture onto a block wall. Dampen the wall first and work the mix well into the pores to help smooth and hide them in the finished job.

Immediately, before the bagging-in mix has time to set, begin the finishing operation. A sponge rubber float can be worked over the blocks to fill in some of the normal surface texturing.

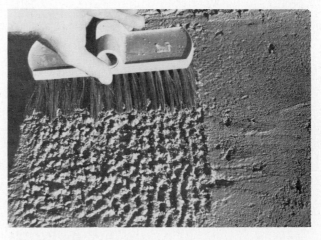

Old, original bagged-in finish is made by rubbing the wall with a damp piece of burlap sacking. To hide joints as much as possible, rub straight across them at right angles or in circles.

Brush dabbed into the fresh cement-plaster coating creates a stippled texture. Once you have the mortar on the wall, it's easy to create other textures. Experiment with different patterns.

Imitation Brick and Stone

Some of these new synthetic bricks and stones look and feel so real that they even fool professional masons!

Here's a sampling (left) of imitation masonry units you can buy: (top row, l. to r.)—Eldorado stones; Wards vermiculite fieldstone; Smoked White and Old Chicago Z-Bricks; Wards Rustic White. (bottom row)—Decro-Wall multi-sized stones; Plastronics bricks and stone; and a Roxite brick panel. At the right is a Roxite outside-corner panel in the same pattern. Inside corners are made by cutting.

A NUMBER of manufacturers have come out with the most real looking imitation bricks and stones that you could ask for. They claim that even a bricklayer can't tell the difference. One manufacturer displays a sample laid-up panel of his bricks containing one real brick and challenging the buyer to find it. Everyone picks the wrong one.

The fake brick and stone units even *feel* real, not soft and yielding like the older vacuum-formed plastic "masonry panels". The truth be known, about the only way you can tell the difference between the artificial and the real is to step up and pick at one of the "mortar" joins with a fingernail. The fake joints, unless they're real-mortar ones, are soft and not always gritty.

A much surer tip-off is the places *where* you see artificial brick and stone masonry used. Because it is very light and needs no footing, the stuff can be put on any interior wall, even going up from wainscot height. That's an impossibility with real masonry.

What's more, fake masonry—the fireproof kind—can be installed right up to a fireplace opening (but not inside the fireboxes). It can go reasonably closely behind a room fireplace. My own Masonite *Roxite* wall comes within 12 inches of a double-insulated Franklin fireplace. Red-hot fires have made the Roxite panels too hot to touch, yet they do fine. The fireproof fakes can be installed even closer to heat. Actually the closest any such wall should come to a fire (without additional protection) is ten inches. The combustible wall behind it is the limiting factor, not the brick.

Indoor-outdoor imitation masonry can be used in living rooms, kitchens, bathrooms, around showers and bathtubs (if rated for outdoor exposure and sealed), basements, family rooms, whatever. And that's only inside the house. The outdoor type can be used for house-remodeling, making a brick or stone planter box, or for building a pair of masonry entrance columns at the front gate. Exterior plywood panels should be used as backing on outdoor projects. Indoors, the existing walls will do.

Cleaning the in-place units is easy with soap and water. The units are thin and intrude very little into the room. A narrow wood molding will cover the exposed edge. although the courses of these artificial bricks and stones are laid perfectly level, you can accommodate an

Above, a living room wall finished with Eldorado antique-style bricks looks and feels so real few people can tell it's not.

Fireproof fake masonry, like Eldorado stone, can be laid right up to a fireplace opening. The fireplace is of regular masonry.

Below, Eldorado masonry units use the mortar-bedding method of installation. Real portland cement mortar is mixed and buttered onto the back of each brick. The process gives a real look.

Left, the wall for a cement-bedded installation of Eldorado bricks is prepared by nailing metal lath to the rough wall. Plastic sheeting laid on the floor protects it from mortar droppings.

The real fun part of the job comes when you press mortared bricks onto the mesh-covered wall. To keep the courses straight, it would help to have chalk level lines snapped at intervals on the wall.

Once the bricks are in place, all that's left is to point the joints. To do it, put some mortar in a canvas pointing tube. Squeeze it out by rolling up the tube. Finish by troweling smooth.

A regular brick pointing tool is used to smooth out the mortar just below the surface of the brick to give you a raked joint. The brick thickness varies slightly to produce the authentic look.

off-level floor by installing the first course of bricks vertically and trimming those where the floor is high. It won't be noticed. Whatever you do, follow the directions that come with the units you select. Prices for imitation masonry vary from a bit more than 50 cents a square foot to more than three dollars a square foot for the fanciest fireproof stones that are suitable for outdoor use.

As you might guess, each manufacturer furnishes the materials you need to do an installation with its products. These include adhesive and grout in mortar colors: gray, white, black; inside and outside corners, where necessary; touch-up paints; and sealers for walls installed in wet locations. Patterns vary almost as much as those available in real bricks and stones, including some style called "Old Chicago". An-

MAJOR IMITATION MASONRY MATERIALS

BRAND	TYPES	MATERIAL	OUT DOOR	HOW INSTALLED	REMARKS
MASONITE ROXITE PANELS	BRICK, STONE 3 STYLES 11 COLOR	CRUSHED LIME STONE	NO	MOUNT WITH NAILS WALL ANCHORS	3 1/3 SQ.FT. PANELS GO UP FAST AND NEAT- CORNERS AVAILABLE HEAT RESISTANT
Z – BRICK	BRICK, STONE 5 STYLES 12 COLOR	VERMICULITE	YES	MOUNT IN ADHESIVE THAT LOOKS LIKE MORTAR	REASONABLE COST. COLOR INTEGRAL WIDELY SOLD FIREPROOF
MONTGOMERY-WARD WHITE BRICK REPLIC.	BRICK ONE STYLE	POLYSTYRENE	NO	SAME AS ABOVE	REAL APPEARANCE VERY LOW COST COMBUSTIBLE 135° MAX. TEMP.
STANDARD BRICK	BRICK, 3 STYLE COLOR	VERMICULITE	NO	SAME AS ABOVE	REAL LOOKING MED. PRICE FIREPROOF
RUSTIC BRICK AND FIELDSTONE	2 BRICKS 2 STONES	VERMICULITE	YES	SAME AS ABOVE	USE ANYWHERE- VARY IN THICKNESS REALISTIC FIREPROOF
SEARS ROEBUCK INDEPENDENCE	RED BRICK	POLYSTYRENE	NO	SAME AS ABOVE	REALISM -LOW COST-COMBUSTIBLE
LEXINGTON GETTYSBURG	BRICK 3 STYLES STONE 1 STYLE	POLYSTYRENE PLASTIC	NO	SAME AS ABOVE	MORE COST - SAND BLASTED FOR NATURAL LOOK - COMBUSTIBLE
CONCORD	BRICK 1 STYLE	CERAMIC	YES	SAME FOR INDOOR- USE CEMENT-MORTAR OUTDOOR	USE ANYWHERE EXCEPT FLOORS FROM JAPAN FIREPROOF
DECRO-WALL BRICK-CRAFT	BRICK 2 COLOR	POLYSTYRENE	NO	STICK ON WITH MOUNT TABS-SQUEEZE ON JOINT GROUT FOR WALL ADHERE	LIGHTWEIGHT- SIMPLEST TO INSTALL VARIED SIZE 'STONE' INTERLAY COMBUSTIBLE - O.K. TO 160°
ELDORADO STONE	BRICK, STONE 9 STYLES 24 COLORS	SHALE PERLITE CONCRETE	YES	SET IN CEMENT MORTAR OVER METAL LATH	DYED COLORS- STONE THICKNESS VARY FOR REAL LOOK - FIREPROOF
PLASTRONICS BRICOVER	BRICK, 4 STYLES 1 FLOOR, STONE 1 TYPE	BRICK PARTI-CLES IN PLAST.	YES	MOUNT IN ADHESIVE WITH GROUTED JOINTS	BRICK WITH OUTDOOR CAP. ONLY ONE SUITABLE FOR FLOOR, HEAT RESIST

Start a Masonite Roxite paneled installation by snapping a level chalkline onto the wall to position the first course. The wall need not be finished or painted; just so it's sound enough to hold.

1. To install Roxite panels, you have a choice of mastic or anchoring. On hollow walls anchoring them is less messy. Position the panel, then drill a hole through it and the wall material.

tique and "used-brick" styles are also available. Your best bet is to see what your dealer has to offer.

INSTALLATIONS

Basically, two installation systems are in use: Adhesive mastic and cement-mortar. With the mastic method, the adhesive is spread like floor tile cement, but thicker, and the units are set in it. A cement-mortar installation is similar, but real mortar is used. With this method, even the joints look real, because they are.

Masonite *Roxite* panels also may be put on with nail and wall anchors, *Decro-wall* is put up with self-stick mounting tabs. Grouting the joints effects a lasting bond to the wall.

One manufacturer—*Plastronics Bricover*—offers a bogus brick style suitable for floor installation.

Difficulty in installing these artificial varies widely among those available. The ones that require a cement-mortar backing are hardest to apply simply because of the mortar-handling that's necessary. They're also the toughest, most fire-resistant and hardest to tell from real.

One of the easiest to put up are the *Roxite*

4. Once the first panel is fastened to the wall, place the next one with the finger-joints interlocked. Top of each panel should touch the chalkline. Floor gap can be covered with a molding.

panels. Instead of installing and aligning just one unit at a time, you put up a whole panel containing perhaps 12 units. This makes the job go much faster. What's more, you can forget about alignment problems among the panelized units. The manufacturer has taken care of that. Just be sure that the panels fit correctly and

2. Insert a special plastic Masonite drive-nail anchor into the drilled hole. Also fasten the panel with ring-grooved plasterboard nails into each stud at top, center and bottom of panels.

3. Insert a drive-nail and hammer it home. Install enough anchors and nails so the panels will hold firm when pressed. The panels are very tough. And if you miss the nail, they'll survive.

5. Keep mounting the 12-brick sections from the bottom up, working from right to left. This wall was ordinary ½" sheetrock with the joints taped. Make certain sections are level.

6. After the wall is completely covered, the joints where each section dovetails with the next is grouted, using a special grout and a calking gun. This calls for strong fingers.

you've got it whipped.

It may break your heart to come to the end of a wall and find that most of a panel must be trimmed off to complete a course. Save the trim-off. Most can be used—with further trimming—to begin the next higher course.

The only drawback in making a panelized Roxite installation was that the joints had to be grouted afterward. This is true with some of the single-unit-materials, too. The mortar is real-looking and it goes on perfectly. However, I found it somewhat hard to force them from the cartridge. For this reason, you'll want to tackle grouting in easy stages.

INDEX